PHYSICS PROBLEMS FOR PROGRAMMABLE CALCULATORS

MECHANICS AND ELECTROMAGNETISM

J. RICHARD CHRISTMAN
U.S. Coast Guard Academy

JOHN WILEY & SONS   NEW YORK
CHICHESTER   BRISBANE   TORONTO

Copyright © 1981 by John Wiley & Sons, Inc.

All rights reserved.

Reproduction or translation of any part of this work beyond that permitted by Sections 107 or 108 of the 1976 United States Copyright Act without the permission of the copyright owner is unlawful. Requests for permission or further information should be addressed to the Permissions Department, John Wiley & Sons, Inc.

ISBN 0 471 08212 0

Printed in the United States of America

10 9 8 7 6 5 4 3 2 1

This book is dedicated to

JOHN D. CHRISTMAN

PREFACE

Use of programmable calculators and computers in the class room and laboratory add a great deal to a physics course. Now that such instruments have become inexpensive and widely available, we believe more and more instructors will incorporate their use into both introductory and upper level physics courses. The purpose of this book is to provide material, in the form of program flow charts and problems, which can be used, along with the programmable machines themselves, to augment the more usual physics courses.

This book is intended to be used as an adjunct to a calculus based introductory course in physics of the type usually taken by engineering and science students. It is helpful, but not necessary, for the students to take a course in computer programming concurrently.

The structure of this book follows the usual order of topics in mechanics and electromagnetism found in most introductory courses. Individual chapters are introduced by reference to material in Halliday and Resnick's books PHYSICS (3rd edition) and FUNDAMENTALS OF PHYSICS (2nd edition), both published by John Wiley & Sons. This material, or similar material in another text, should be studied before attempting the calculator problems.

No physics beyond that usually covered in an introductory course is used. The programs illustrate applications of Newton's second law, Coulomb's law, the Biot-Savart law, Gauss's law, Faraday's law, and Kirchoff's rules. Many of the problems are meant to illuminate the meaning of these laws through computation. To put the programs into context, some discussion of physical principles is included, but this is not meant to substitute for the discussions in the course text.

As is appropriate for a book which contains calculator applications, the physical situations presented in the problems are often computationally more difficult than those found in the usual introductory text. The mathematical processes used, however, are always closely related to those used in introductory courses to obtain analytic solutions to simpler problems. All numerical techniques used are fully discussed.

This book can be used in several different ways. In some cases, the actual programming as well as the shorter exercises and problems can be assigned as homework and discussed in class. Since many of the problems are associated with standard laboratory experiments, some of them can be included as part of the laboratory work. The longer problems can be associated with study projects for interested students. They can be included, for example, in honors courses.

It is not expected that all the material in this book can be completed in addition to the material in the course text. It is possible, however, to inject more than a few calculator topics into the introductory course. A great variety of topics is available to choose from. All, we hope, suggest numerous additional problems and study projects which can be tailored to fit a wide variety of course structures and time periods.

The programs in this book have been designed to run on the better hand held programmable calculators as well as on larger computers and, in order to work the problems, the reader must have access to one of these machines. Though some programs require larger memory than is available with the smaller capacity hand held machines, the overwhelming majority of the problems and exercises can be solved no matter which programmable machine is used. A few of the problems make use of indirect addressing of memory locations and it is helpful, but not necessary, to have access to a machine with this capability. When working the longer problems, it is also helpful to use a machine which prints results.

Most of the problems and exercises take less than half an hour to run on a hand held machine. A few, most notably those dealing with two dimensional motion and the plotting of field lines, take considerably longer. It is worthwhile to use a mini-computer or a large machine for these problems.

Programs are given as flow charts, using a language which is similar to what is referred to as BASIC. For most programmable machines, the given flow chart statements need to be translated into actual computational steps and each line in a flow chart may actually represent a large number of such steps. To give these steps for each program and to give them over again, once for each of the four or five most common programming languages, would make this book extremely large and hard to use. The flow chart, on the other hand, does give the essentials of the program and appendices have been included to help the student translate the flow charts into actual program steps. With this help, and some study of the instruction manual for the machine used, the student should have little difficulty making the translation.

The programs given in this book are not necessarily optimum. Rather, simplicity has been a goal in writing them. Students with some experience in programming can make use of the capabilities of particular machines to produce programs which are faster and more efficient than those presented here.

Data given in most of the problems have been selected so that the answers are correct to 3 significant figures, provided the machine is programmed and run correctly. If the results are used primarily to plot graphs, the accuracy is reduced to 2 figures to improve running time. When this occurs, mention is made of how data can be modified to improve accuracy. We mean all input data to have at least 3 figure accuracy, although fewer figures may actually be given.

It is hoped that students, reading and working with this book, will sense the enormous range of applicability of the laws of physics and will come to realize a higher level of understanding of those laws.

For their contribution of many useful discussions and for their support, the author thanks Saul Krasner, Gregory Cope, Bruce Russell, William Helgeson, Robert O'Hara, and Michael Bray. Thanks also goes to Ruth Pflomm, who typed the manuscript. The author is indebted to Robert McConnin and Francine Fielding of Wiley, who helped in innumerable ways.

The author wishes to express his gratitude to Mary Ellen, Stephen, and Karen Christman, and Ellen Tossett, who helped with the manuscript and who exhibited enormous patience and good spirit during the writing of this book. Their support and encouragement are greatly appreciated.

November 1980
New London, Connecticut 06320

J. RICHARD CHRISTMAN
U. S. Coast Guard Academy

TABLE OF CONTENTS

Chapter 1   PROGRAMMING   1

1.1   Introduction   1
1.2   The Language of Programming   2
1.3   Programming   6
1.4   Flow Charts   9
1.5   Exercises   13

Chapter 2   PROGRAMS FOR FINDING ROOTS AND FOR PLOTTING   16

2.1   Why Find Roots?   16
2.2   The Method of Uniform Intervals   16
2.3   Binary Search Method   20
2.4   A Plotting Program   23
2.5   Exercises   25

Chapter 3   MOTION IN ONE DIMENSION   27

3.1   Time as the Root of Kinematic Functions   27
3.2   Parameters of the Motion as Roots   29
3.3   Exercises   31
3.4   Motion in a Uniform Gravitational Field with a Drag Force   32

Chapter 4   CURVE FITTING   37

4.1   The Fourth Order Polynomial   37
4.2   Improved Efficiency   40
4.3   Use of the Polynomial for Interpolation and Extrapolation   42
4.4   Use of the Polynomial for Differentiation   46
4.5   Exercises   48

Chapter 5   MOTION IN TWO DIMENSIONS   51

5.1   Projectile Motion   51
5.2   Projectile Motion Exercises   55
5.3   Uniform Circular Motion   57

Chapter 6   INTEGRATION OF NEWTON'S SECOND LAW WITH APPLICATION TO MOTION IN ONE DIMENSION   61

6.1   The Basic Program   61
6.2   Velocity Independent Forces   65
6.3   Tests of Accuracy   69
6.4   Exercises   72

Chapter 7   NEWTON'S SECOND LAW IN TWO DIMENSIONS   75

7.1   Programs for Two Dimensional Motion   75
7.2   Projectile Motion with a Drag Force   79
7.3   Circular and Nearly Circular Orbits   83

Chapter 8   INTEGRATION, WITH APPLICATIONS TO THE
            CALCULATION OF WORK   86

8.1   Integration by Simpson's Rule   86
8.2   Calculation of Work for Motion in One Dimension   90
8.3   Time and Velocity Dependent Forces   92
8.4   Work for Motion in Two Dimensions   94
8.5   Exercises   98
8.6   Potential Energy Functions   100

Chapter 9   ROTATIONAL MOTION   103

9.1   Rotational Kinematics for Motion about a Fixed Axis   103
9.2   Rotational Dynamics for Motion about a Fixed Axis   105
9.3   Exercises   107
9.4   Conservation of Angular Momentum   110

Chapter 10   PHYSICS OF SPECIAL SYSTEMS   115

10.1   Oscillators   115
10.2   Motion in a Gravitational Field   120
10.3   Rocket Motion   128

Chapter 11   THE ELECTRIC FIELD   132

11.1   The Electric Field of Point Charges   132
11.2   Continuous Distributions of Charge along a Line   139
11.3   Exercises   144

Chapter 12   ELECTRIC POTENTIAL, FIELD LINES, AND GAUSS'S LAW   149

12.1   Electric Potential   149
12.2   Field Lines   158
12.3   Gauss's Law   163

Chapter 13   SOLUTION OF SIMULTANEOUS EQUATIONS
             WITH APPLICATION TO CIRCUIT PROBLEMS   171

13.1   Simultaneous Equations   171
13.2   Exercises   180
13.3   Direct Current Circuits   181

Chapter 14   THE MAGNETIC FIELD   189

14.1   Magnetic Fields of Moving Charges   189
14.2   Magnetic Fields of Circular Current Loops   196
14.3   Ampere's Law   205

Chapter 15   MOTION IN ELECTRIC AND MAGNETIC FIELDS   210

15.1   The Force Equations   210
15.2   Motion in an Electric Field   213
15.3   Motion in a Magnetic Field   215
15.4   Motion in Crossed Electric and Magnetic Fields   218
15.5   Magnetic Forces and Torques on Circuits   219

Chapter 16   TIME DEPENDENT MAGNETIC FIELDS   226

16.1   Faraday's Law   226
16.2   Circuits with Inductance   227

APPENDICES   241

A.   Texas Instruments Machines   243
B.   Hewlett-Packard Hand Held Machines   254
C.   BASIC   266
D.   FORTRAN   276
E.   PASCAL   288

TABLE OF PROGRAMS

Vector addition.	12
Method of uniform intervals for finding roots.	18
Binary search method for finding roots.	22
Evaluation of a function for plotting.	24
Evaluation of the coefficients for a fourth order polynomial. Interpolation and extrapolation.	44
Modification of polynomial fit program for a function in tabular form.	45
Evaluation of the derivative using a polynomial fit.	47
Modification of the binary search program to accept polynomial coefficients.	47
Integration of Newton's second law in one dimension. General force.	64
Integration of Newton's second law in one dimension. Velocity independent force.	68
Integration of Newton's second law in two dimensions. General force.	76
Integration of Newton's second law in two dimensions. Velocity independent force.	78
Integration of a function by Simpson's rule.	89
Work done by a force for motion in two dimensions.	97
Newton's second law programs for rotation about a fixed axis.	106
Work done by a torque.	107
Integration of Newton's second law for a gravitational force.	123
Inclusion of energy and angular momentum calculations in Newton's second law programs.	125
The electric field of a collection of stationary charges. Charge information is entered for each field point.	134
The electric field of a collection of stationary charges. Charge information is stored.	137

Modification of Simpson's rule program to calculate the electric field of a continuous line of charge.	141
The electric field of a continuous line of charge with slowly varying charge density.	143
The electric potential of a collection of stationary charges.	152
The electric potential of a continuous line of charge.	153
Calculation of points along electric field lines.	160
Calculation of the electric flux through the face of a cube.	168
Solution of simultaneous linear algebraic equations.	175
The magnetic field of two moving charges.	191
Calculation of points along magnetic field lines.	194
The magnetic field of a circular current loop.	200
Evaluation of the line integral in Ampere's law.	207
Calculation of the current in circuits with self-inductance.	231
Calculation of currents in inductively coupled circuits.	239

PHYSICS PROBLEMS FOR
PROGRAMMABLE CALCULATORS

Mechanics and
Electromagnetism

Chapter 1

PROGRAMMING

In this chapter we introduce the language used throughout this book to write calculator programs. The last example and the exercises are problems in vector analysis. These should be worked in conjunction with Chapter 2 of PHYSICS, Chapter 2 of FUNDAMENTALS OF PHYSICS, or an equivalent chapter in another text.

1.1 <u>Introduction</u>

All programmable calculators have the capability of performing certain basic functions. They receive data (usually from a keyboard), store data, retrieve data from storage, perform arithmetic and other numerical operations on data, and display or print results. All this is done according to some sequence of instructions, called a program, which the operator stores in the memory of the calculator and which the calculator executes.

Allowed instructions correspond to the basic functions mentioned above. The most important numerical operations include addition, subtraction, multiplication, division, exponentiation, and the evaluation of certain useful functions such as the sine and arc sine, cosine and arc cosine, tangent and arc tangent, square and square root, exponential and natural logarithm. Evaluation of other functions are possible on some calculators.

Unless told to do otherwise, the calculator always executes the instructions in the order in which they are given to it in the program. When the machine is finished executing an instruction, it normally proceeds to the next one in the sequence. There are, however, certain instructions which cause the machine to jump forward or backward to some predetermined place in the program and start executing instructions in sequence there. These are called transfer instructions and they are responsible for a great deal of the flexibility of programmable calculators.

In the rest of this chapter we discuss how we write instructions in this book and give enough of the rudiments of programming that you should be able to follow and understand the programs we present later on.

1.2 The Language of Programming

It is not the purpose of this book to present keystroke by keystroke listings of any program, since to do so would narrow its usefulness considerably. Programs are given, however, in a language which is easy to translate into a sequence of keystrokes for almost any programmable calculator.

The language we use is similar to the common programming language called BASIC. Because we use only those parts of the language which describe operations nearly all programmable calculators can perform, it is easy, for each program, to list the keystrokes which cause the calculator to execute the program. The reader is encouraged to do this for each program statement in this chapter until the translation becomes second nature. Appendices, one for each of several widely used languages, are included to help make the translation. For each programming statement given in this book, find a similar statement in the appendix appropriate for your machine. There the keystrokes are given.

We now proceed to describe the programming language used throughout this book.

A calculator has a number of data storage locations in its memory; each location can hold one number. Stored numbers can be recalled from memory, then used in numerical operations, and the results of numerical operations can be stored in memory. For most calculators, the various storage locations are identified by numbers. We use the symbols $X1$, $X2$, $X3$, ... to represent storage locations 1, 2, 3, ... respectively.

We make great use of these storage location symbols. They form the backbone of the language and appear in nearly all program statements. As will become apparent, most of the time any of these symbols can also be interpreted as the number stored in that location. It is not wrong and it is usually convenient to think of $X1$, for example, as representing the number stored in location 1.

A physical quantity is often associated with a specific storage location. For example, it may be that the y component of the position vector for a particle is always stored in memory 1. We can then take $X1$ to be the algebraic symbol for that physical quantity. For clarity in writing and ease in understanding a program, it is worthwhile to retain the association between physical quantity and storage location throughout a program. Whenever possible, this is done for the programs presented in this book. As soon as possible, you should acquire the habit of thinking of X_n as the storage location, the number in that location, and the physical quantity.

The four arithmetic operations are written

$$X_n + X_p$$
$$X_n - X_p$$
$$X_n * X_p$$
$$X_n / X_p$$

The first of these means add the number stored in X_p to the number stored in X_n. The meaning of the others is just as clear. Note the symbol * is used to denote multiplication to avoid confusion with the letter x. Exponentiation is written

$$X_n \uparrow X_p$$

which means: raise the number in X_n to the power given by the number in X_p.

The most widely recurring functions are written

SIN(X_n)	INVSIN(X_n)
COS(X_n)	INVCOS(X_n)
TAN(X_n)	INVTAN(X_n)
EXP(X_n)	LN(X_n)
SQRT(X_n)	ABS(X_n)

There should be no trouble interpreting most of these symbols. INV is used for the inverse trigonometric functions. The last operation is used to find the absolute magnitude of the number in X_n. By convention, we take all angles to be in radians unless some other unit is specified.

For purposes of this book, all instructions which contain numerical operations also give a storage location for the result of the operations. The form taken by all such instructions is

$$\text{numerical operation} \rightarrow X_n$$

which is interpreted to mean: perform the numerical operation indicated on the left side of the arrow and store the result in Xn.

An example is

$$X1 + X2 \rightarrow X3$$

This instruction tells the calculator to add the contents of X1 and X2, then store the result in X3. For most calculators, the contents of X1 and X2 must first be recalled from memory to registers where the arithmetic is performed, and a specific instruction is usually required to store the result. The contents of X1 and X2 are not altered by either the arithmetic or the storage instruction. If a number is in X3 before the storage instruction is executed, it is erased and replaced by the result of the operation.

Instructions can be more complicated than the one given above. Some examples are:

$$X1 * SIN (X2 + X3) \rightarrow X4$$
$$INVTAN (X1 * SIN (X2)) \rightarrow X3$$
$$SQRT (ABS (X1 / X2)) \rightarrow X3$$

Each of the above instructions leaves unaltered the numbers which are stored in the memory locations mentioned on the left side of the arrow.

It is sometimes useful to write instructions which do alter the content of one of the memory locations. One such instruction is

$$X1 + X2 \rightarrow X1$$

In this case, the contents of X1 and X2 are added and the sum is placed in X1. The original content of X1 is erased in the process and it is clear that this type instruction can be used only if the original number in X1 is not required by the remainder of the program. Use of this type instruction saves memory locations. In many cases, it also makes the program more succinct and easier to read. It preserves the identification between memory location and physical quantity.

By convention, we assume an order to the processing of arithmetic operations in the same statement. Exponentiation is carried out first, then multiplication and division, and finally addition and subtraction. As examples, the multiplication in X1 + X2 * X3 is carried out before the addition and the exponentiation in X1 * X2 ↑ 2 is carried out before the multiplication. In the first case, it is the product of X2 and X3 which is added to X1 and, in the second case, it is the square of X2 which is multiplied by X1.

Parentheses are used freely. The inner most set is evaluated first, then the next inner most and so forth until the outer most set has been evaluated. Within each set of parentheses, arithmetic operations are carried out in the order described above.

As mentioned previously, a numerical quantity can be identified by the memory location where it is stored. We use this idea when we write the data entry and display instructions as

    ENTER Xn

and   DISPLAY Xn

respectively. The first of these instructs the machine to accept a number from the keyboard and store whatever number it receives in Xn. The second tells the machine to display the number stored in Xn.

From the point of view of efficiency, it is not always desirable to store numbers immediately. Most calculators allow operations to be performed on an entered number while it is still in the display register. Nevertheless, instructions in this book are written with entering numbers going straight to memory. This way of writing retains the identification of a number with its storage location.

Most machines allow the use of scientific notation (powers of 10) for entering and displaying numbers. When needed, we also use such notation. The power of 10 follows the letter E (for exponent). For example 1.075E-12 means 1.075×10^{-12}.

1.3 Programming

A calculator program consists of a sequence of instructions which the machine executes (in order, except when instructed to do otherwise by one of the instructions). A few examples should suffice to show how programs are constructed. These examples also introduce some of the conventions used in this book.

Suppose we wish to evaluate the function $y=0.3\sin(0.02t^2)$ for various values of the independent variable t. We intend to enter the values of t from the keyboard. A program which does the job is

```
ENTER X1
.3 * SIN (.02 * X1 ↑ 2) → X2
DISPLAY X2
STOP
```

Here the value of t is entered and stored in X1, the appropriate numerical operation is performed with the result placed in X2, the number is X2 is displayed, and the machine is told to stop.

This program evaluates y for only one value of t. To evaluate y for more values of t, we want a program which returns to the beginning to accept a new value of t. This requires the use of what is known as an unconditional transfer statement or GOTO statement. First we label the first statement of the program with an A, thusly:

```
"A"; ENTER X1
```

Then, after STOP, we add the transfer statement

```
GOTO A
```

When the calculator reaches this instruction, it searches for the statement labelled A and executes it next. The complete program is

```
"A"; ENTER X1
.3 * SIN (.02 * X1 ↑ 2) → X2
DISPLAY X2
STOP
GOTO A
```

The STOP instruction is retained so that the operator can write down the displayed number. When the machine is restarted, it proceeds to the next instruction, GOTO A. In some programming languages, statements are labelled by means of numbers rather than letters.

In order to give an example of a slightly more complicated program, assume we wish to evaluate y for a sequence of t values at equal intervals Δt. We let X1 be the value of t, X2 be the interval Δt, and X3 be the value of y. The program is

```
ENTER X1
ENTER X2
"A"; .3 * SIN (.02 * X1 ↑ 2) → X3
DISPLAY X3
STOP
X1 + X2 → X1
GOTO A
```

To start the program, we enter the initial value of t and then the interval width. The function is evaluated at that value of t, the result is displayed, and the machine stops to allow the operator to copy the result. When restarted, the machine adds the interval width Δt (in X2) to the current value of t (in X1). This produces the next value of t and it is stored in X1. The machine then returns to statement A where it evaluates the function for the new value of t.

The group of statements including the one labelled A, the GOTO, and all those between are collectively called a loop. The machine executes the sequence of instructions in the loop repeatedly (unless, of course, the operator fails to restart the machine after a display). The loop is an important device for programming when the same operations must be repeated many times. Nearly all programs in this book contain loops.

For the program above, the machine stops every time a new value of y is displayed, then goes on to compute the next value of t when it is restarted. The machine requires operator attention. If the machine has printing capability, we can avoid this by asking the machine to print the result. Then, of course, there is no need for the operator to copy the value of y. Replace the DISPLAY instruction with PRINT X3 and delete the STOP instruction.

A new problem now arises. The machine might calculate and print more values of y than are desired, perhaps using a large quantity of paper. The solution is to use a conditional transfer or IF statement.

We give an example of an IF statement which will get the machine out of the loop. Suppose we wish only the first 8 values of y. We then program the machine to count the number of times it starts the loop. We need an instruction so that when this number is 9, the machine goes to an instruction outside the loop rather than executing the statements of the loop another time. The appropriate statement is

$$\text{IF } X4 = 9; \text{ GOTO B}$$

Here X4 is the number of times the machine starts through the loop and the instruction means: if X4 is not 9, the machine next executes the statement following the IF statement, just as if GOTO B were not there. If, on the other hand, X4=9, then the machine next executes the instruction labelled B. Since the job is done when X4=9, statement B is a STOP statement.

The complete program is

```
ENTER X1
ENTER X2
0 → X4
"A"; X4 + 1 → X4
IF X4 = 9; GOTO B
.3 * SIN (.02 * X1 ↑ 2) → X3
PRINT X3
X1 + X2 → X1
GOTO A
"B"; STOP
```

Notice that the program must set X4 equal to 0 before the loop is entered. This is because, at A, X4 is incremented by 1 to count the number of times the loop is started.

X4 must have an initial value so that there is a number to add to 1 and the initial value must be 0 so that the first time the loop is started, X4 is incremented to 1.

There are several other forms of conditional transfer statements. Some are

and
$$\text{IF } X1 < X2;\ \text{GOTO A}$$
$$\text{IF } X1 > X2;\ \text{GOTO A}$$

In each case, if the inequality is true, then the machine goes to the statement labelled A. If the inequality is false, then the machine continues as if the transfer statement were not there.

1.4 Flow Charts

A flow chart is a diagram which, for a particular program, gives all the numerical operations and shows the order in which the operations are carried out. It also shows ENTER, DISPLAY, IF, and STOP statements. The flow chart for the last program of section 1.2 is shown in Fig. 1-1.

Each statement is in a box: numerical instructions and the DISPLAY instructions are in rectangular boxes, the ENTER instruction is in a rectangular box with its corner cut off, the IF statement is in a diamond shaped box, and the STOP statement is in a circular box.

The boxes are numbered so that discussions of the program can refer to them individually. Although the boxes are numbered in order of appearance, the numbers used are not consecutive. Numbers are omitted to allow for other statements which might be added if the program is modified.

The lines connecting the boxes give the order of execution of the instructions. In running the program, the machine follows the arrows from one instruction to the next. At the IF statement, line 130, the machine is asked to test the equality: is X4 equal to 9? If it is not, the program proceeds straight through the diamond shaped box to line 140. If the equality is true, the program changes direction and goes to the STOP instruction at line 200.

Figure 1-1. Flow chart for the last program of section 1.2.

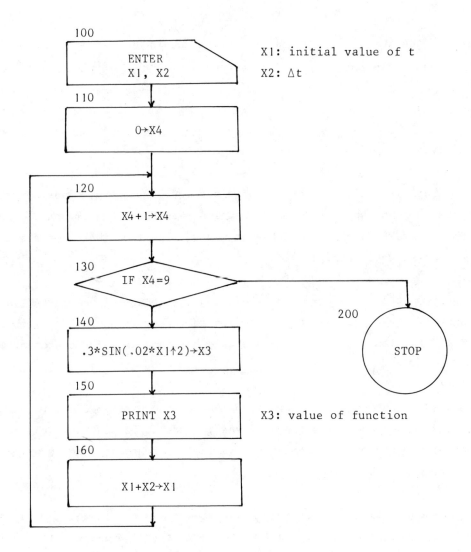

The unconditional transfer statement is not explicitly shown on the flow chart. It is represented by the flow line from the last statement in the loop, line 160, back to just above the first, line 120. The appearance of flow lines which indicate a change in direction for the program are signals that transfer instructions must be written and instructions must be labelled when the flow chart is translated into machine instructions. There is an implied GOTO instruction following line 160 and lines 120 and 200 must be labelled.

Several conventions are used in the flow charts of this book. One of them has already been used in the flow chart above. When there are several numbers to be entered in succession, we list them all in the same ENTER statement rather than write a separate statement for each. The operator must remember that ENTER X1, X2 may stand for two separate "enter" keystrokes (as well as two keystrokes to place the entered numbers in memory).

The same convention is used for DISPLAY and PRINT statements. DISPLAY X1, X2 is used instead of separate statements for each number, for example. This also implies a STOP instruction after each separate display keystroke if the operator wishes to copy the displayed number.

For some flow charts in this book, two or more numerical operation instructions may be placed in the same box. Instructions in the same box are always related to each other in some way and the grouping helps to highlight the connection as well as saves space.

For many of the programs, the user must supply a function. For example, the same program can be used for many different mechanics problems, for which the force acting is different, and the user must supply programming statements so that the machine can evaluate the force for a particular problem.

Suppose, for example, the quantity we need to evaluate is a function of two independent variables and we store one of them in X1 and the other in X2. As part of a flow chart, we write

$$f(X1,X2) \rightarrow X3$$

It means: use the numbers in X1 and X2 to evaluate the function and store the result in X3. In actual use, the operator must supply the correct line which gives the machine the exact numerical operations it needs to evaluate the function.

As another example of a program, we consider the addition of two vectors $\vec{A}$ and $\vec{B}$. If the cartesian components of the vectors are given, then the cartesian components of their sum $\vec{C}=\vec{A}+\vec{B}$ are found using

$$C_x = A_x + B_x,$$
$$C_y = A_y + B_y,$$
and
$$C_z = A_z + B_z. \tag{1-1}$$

A flow chart is shown in Fig. 1-2.

Figure 1-2. Flow chart for a program to calculate the cartesian components of the sum of two vectors.

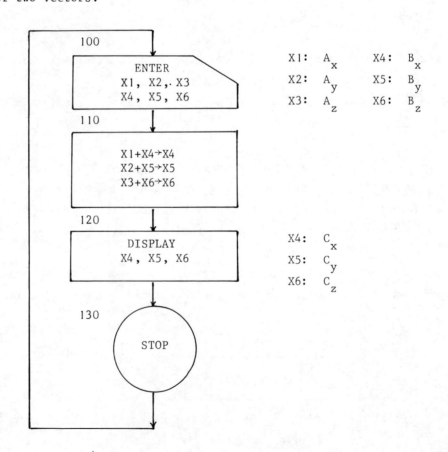

Storage locations are

X1	A_x,
X2	A_y,
X3	A_z,
X4	B_x (then C_x),
X5	B_y (then C_y),
X6	B_z (then C_z).

and

At line 100, the six components are entered and stored. At line 110, the three components of $\vec{C}$ are found and stored, in X4, X5, and X6 respectively. Since the components of $\vec{B}$ are no longer needed, these steps are valid and, furthermore, they save some storage space in the machine. The components of $\vec{C}$ are displayed at line 120. The machine stops to allow them to be copied and then, when it is restarted, it returns to line 100 to consider two new vectors. This program can be used as a model to work some of the following exercises.

1.5 Exercises

1. a. Translate the flow chart of Fig. 1-1 into keystrokes for your machine and program the machine. The argument of the sine function is in radians.
    b. Use the program to calculate $y(t)$ for values of t from t=0 through t=9. Increment t by 1 each time around the loop. So that you can check your program, the answers are given below.

t	y(t)	t	y(t)
0	0	5	0.144
1	0.00600	6	0.198
2	0.0240	7	0.249
3	0.0537	8	0.287
4	0.0944	9	0.300

    c. Reprogram the machine to evaluate the function $y(t)=(t^2+0.6)^{3/2}$, given the initial value of t and the interval width. Use the program to find $y(t)$ for values of t from t=0 through t=9 with an interval width of 1.

2. Translate the flow chart for the vector addition program into keystrokes for your

machine, program the machine, and use the program to find the following vector sums.
a. $(5\hat{i} - 10\hat{j} + 25\hat{k}) + (-14\hat{i} + 3.7\hat{j} - 13.3\hat{k})$
b. $(14\hat{i} + 6\hat{k}) + (-7.5\hat{j} + 3.2\hat{k})$
c. $(\hat{i} + 7.1\hat{j}) + (3\hat{i} - 17.4\hat{k})$

3. Design a flow chart for a program to evaluate the scalar product of two vectors, given their cartesian components. Have the machine return to the beginning to consider two new vectors when it is finished with the evaluation. Translate the flow chart into keystrokes for your machine, program the machine, and use the program to evaluate the following scalar products.
a. $(3\hat{i} + 4\hat{j} + 3\hat{k}) \cdot (\hat{i} + 3\hat{j} + 2\hat{k})$
b. $(3\hat{i}) \cdot (7\hat{i} - 3\hat{j})$
c. $(5.7\hat{i} - 6.2\hat{j} + 2.1\hat{k}) \cdot (2.3\hat{i} + 7.6\hat{j} + 3.8\hat{k})$

4. The magnitude of a vector can be found by taking the square root of the scalar product of the vector with itself. Design a flow chart for a program which evaluates the magnitude of a vector, given its cartesian components. Have the machine return to the beginning to consider a new vector when it finishes with the evaluation. Translate the flow chart into keystrokes for your machine, program the machine, and use the program to evaluate the magnitudes of the following vectors.
a. $7.1\hat{i} + 13.2\hat{j} - 16.1\hat{k}$
b. $12.6\hat{i} - 7.1\hat{j}$
c. $-5.8\hat{j} + 4.1\hat{k}$

5. The angle between two vectors can be found by evaluating the arc cosine of the scalar product divided by the product of the magnitudes. For example, the angle Θ between $\vec{A}$ and $\vec{B}$ is given by

$$\Theta = \cos^{-1} \frac{\vec{A} \cdot \vec{B}}{AB}$$

Design a flow chart for a program to find the angle between two vectors, given their cartesian components. Program the machine to return to the beginning to consider two new vectors when it has finished with the calculation.

There is more than one angle with the same cosine. One normally requires the one which is positive and less than π radians. Since most machines automatically produce that angle, no special care need be taken.

Translate the flow chart into keystrokes for your machine, program the machine, and use the program to find the angle between the following pairs of vectors.
a. $(7.4\hat{i} + 3.7\hat{j})$, $(3.9\hat{i} + 4.7\hat{j})$
b. $(-3.7\hat{i} + 1.2\hat{j})$, $(3.7\hat{i} + 1.2\hat{j})$
c. $(4.2\hat{i} + 12.1\hat{j} + 7.3\hat{k})$, $(2.3\hat{i} - 7.8\hat{j} - 5.3\hat{k})$

6. The program written for exercise 5 can be used to find the angle a vector makes with one of the cartesian coordinate axes. The second vector of the pair is simply any vector along the selected axis. Use the program of exercise 5 to find the following angles.
    a. The angle between $3\hat{i} - 7\hat{j}$ and the x axis.
    b. The angle between $3\hat{i} - 2\hat{j} + 6\hat{k}$ and the x axis.
    c. The angle between $3\hat{i} - 2\hat{j}$ and the z axis.
    d. The angle between $3\hat{i} - 2\hat{j} + 6\hat{k}$ and the z axis.

7. Use programs developed for previous exercises to do the following.
    a. Find the angle between $7\hat{i} + 9.6\hat{j}$ and the x axis.
    b. Now add $5.2\hat{j}$ to the original vector and find the angle between the x axis and the vector which represents the sum. Did the angle increase or decrease? Draw a vector diagram which shows the two vectors and their sum. Graphically substantiate the numerical result.
    c. Now subtract $5.2\hat{j}$ from the original vector of part a and find the angle between the x axis and the vector which represents the difference. Draw a vector diagram which shows the two vectors and their difference. Graphically substantiate the numerical result.

Chapter 2
PROGRAMS FOR FINDING ROOTS AND FOR PLOTTING

In this chapter, some numerical techniques for finding roots of functions are discussed and flow charts for calculator programs are given. Exercises for practice are suggested at the end of the chapter. These techniques are among the most important for numerical analysis and many practical physical applications appear in later chapters.

2.1 Why Find Roots?

There are a great many physics and engineering problems which require finding the root of a function: given the function $f(t)$, for what values of t does the function vanish? Given the speed of an object as a function of time, at what time does the object come to rest? Given the intensity of light as a function of position along some line, what points are dark? All maxima and minima problems, of course, can be formulated in terms of finding a vanishing derivative.

If the function is particularly simple, it may be possible to solve for its roots by direct algebraic methods. These methods are usually faster than the numerical techniques to be described and they should be used if possible. There are, however, many functions of interest whose roots can be found only by numerical methods. Two such methods, the method of uniform intervals and the binary search method, are discussed in this chapter.

2.2 The Method of Uniform Intervals

To apply this method, the function is evaluated at a number of values of the independent variable and a check is made to see if the function changes sign from one point to a neighboring point. Fig. 2-1 shows the graph of a function $f(t)$ with the t axis divided into 5 intervals of equal width. Between t_3 and t_4 the function changes sign and the product $f(t_3)f(t_4)$ is negative. To find the position of the root more accurately, the region between t_3 and t_4 can be divided into a number of smaller intervals and the process repeated. Always, the signal that a root has been found is that the product of the values of the function at the two ends of an interval is negative.

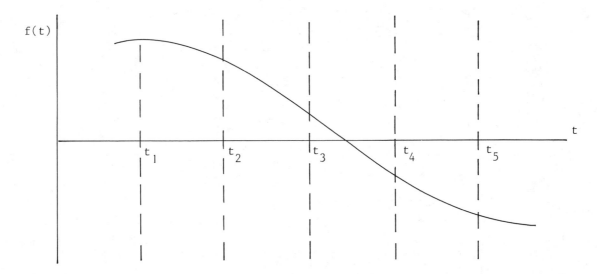

Figure 2-1. The t axis divided into intervals in order to locate the root of f(t).

Fig. 2-2 gives the flow chart for a program which does the job. The initial and final values of t are entered in X1 and X2 respectively and the number of intervals to be searched is stored in X3. At line 110, the width of an interval is calculated and stored in X5. At line 120, the function is evaluated at the first point (t_0) and the value is stored in X7 and, at line 130, the value of t_0 is placed in X8. These last two steps are necessary in preparation for entering the loop at the next statement.

The loop runs from line 200 through line 240 and the program executes the instructions in the loop once for each of the intervals. At the end of the loop, X8 contains the value of t at the right end of the interval being considered, X6 contains the value of the function at the left end, and X7 contains the value of the function at the right end.

At line 200, these storage locations contain numbers appropriate to the previous interval. Since the left end of the interval is the same as the right end of the previous interval, what is required at the beginning of the loop is to transfer the content of X7 to X6, increment X8 by the width of the interval, and evaluate the function at the value of t now stored in X8. This is the value of the function at the right end of the interval and is stored in X7. The reader should follow the flow chart to see how these steps are carried out.

Figure 2-2. Flow chart for the method of uniform intervals.

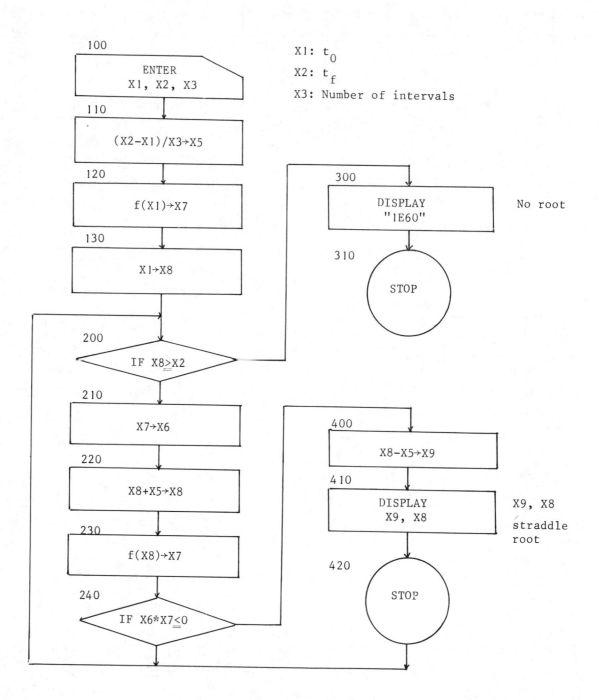

At line 200, a check is made to see if the interval is in the range we wish to consider. If X8 is greater than X2 (the largest value of t considered), then the program does not check for a sign change in the interval but, instead, displays the number 1×10^{60} and stops (see lines 300 and 310). The number 1×10^{60} appearing on the calculator display indicates the program has searched the entire range of t and has not found a zero. This number has been picked as a signal since it is never displayed as the root of any function we consider.

On the other hand, if X8 is less than X2 at line 200, then the interval is checked for a change in the sign of the function. The program passes to line 210, where the number in X7 (the function at the right end of the previous interval) is placed in X6. At 220, X8 is incremented by X5, the interval width, so that X8 contains the value of t at the right end of the interval. This value is then used to evaluate the function at the right end of the interval and this value is placed in X7.

At line 240, the product of X6 and X7 is calculated and its sign is checked. If the product is negative or zero, the program passes to line 400 where the value of t at the left end of the interval is calculated and stored in X9. X8 still contains the value of t at the right end. The two values of t which straddle the root are displayed and the machine stops.

After the contents of X9 and X8 are copied, the program can be restarted and, if it is, it proceeds to line 200 to consider the next interval. This feature is included in the program so that the function can be checked for more than one root.

If, on the other hand, the product of X6 and X7 is positive, the program proceeds to line 200 without stopping. The machine is now ready to look for a change in sign in the next interval. X8 now contains the value of t at the left end of the new interval and X7 contains the value of the function at the left end of the new interval.

Some care must be taken in using this program. If an interval contains an even number of roots, the sign of the function is the same at the two ends and the program will not find the roots. The cure is to use a smaller interval width so that only one root falls in any one interval. On the other hand, running time increases with the number of intervals searched, so it is important to keep the number of intervals from becoming unreasonably large. Some compromise must be struck.

It should be mentioned that there is one class of functions whose roots cannot be found by means of this program. If the function has the same sign on the two sides of the root, as does the function $y=t^2$, then the program will not recognize the root.

Finally, there is a situation for which the program finds a false root. A function can change sign from one side to the other of a point at which it blows up. For example, $\tan(t)$ blows up at $t=\pi/2$ and the sign is positive for angles slightly less and is negative for angles slightly greater. The program produces $\pi/2$ as the root.

As another example, consider the root of $f(t)=At-B$ where A and B are constants. It is clearly $t=B/A$. However, the program might have been asked for the root of $A-B/t$, in which case it produces $t=0$ as well as $t=B/A$.

These problems can be avoided by noticing where the function blows up and omitting those regions from the search or by rewriting the function so that it does not blow up in the range considered.

2.3 Binary Search Method

The program just described runs rather slowly. To obtain high accuracy, it must be run many times with the region searched narrowed each time. Once two points which straddle the root have been found, they can be used as input data for a program which runs much faster. The faster program employs a technique called the binary search method and is described in this section.

Suppose it is known, by means of the uniform interval method or otherwise, that the root of a function lies between t_1 and t_2. The function is then evaluated at the midpoint $(t_1+t_2)/2$ and the signs are checked to determine in which half of the original interval the root lies. Suppose it is to the right of the midpoint. Then the new interval from $(t_1+t_2)/2$ to t_2 is used and the process is repeated. If the root is to the left of the midpoint, the new interval runs from t_1 to $(t_1+t_2)/2$.

The program can be stopped after a predetermined number of halvings. It is usual, however, to stop the program when the interval is smaller than a predetermined size. In this way the user can control the precision with which the root is found.

A flow chart is given in Fig. 2-3. The program is designed to be used alone or in conjunction with the method of uniform intervals. If the machine has a sufficiently large memory, the STOP at line 400 of the previous program is removed and this program added in its place. If this is done, line 500 must be modified to read ENTER X1.

The two points which straddle the root are stored in X8 and X9 respectively. They may be entered from the keyboard or, if the program follows the uniform interval program, nothing need be done since the appropriate values are already in X8 and X9. X1 is the precision with which the root is desired. The machine stops its search when the interval becomes smaller than the number entered into X1.

At lines 510 and 520, the function is evaluated at the two points and the values placed in X2 and X3 respectively. At 530 the midpoint is calculated and, skipping line 540 for the moment, the value of the function at the midpoint is found at line 600 and stored in X5.

At 610, a check is made to see if the function has the same sign at the midpoint as at X8 (the right end of the interval if the output of the uniform interval program is used). If the product is positive, the root must be between the midpoint and X9 and the program goes to line 650. There, the old interval is changed to the new one so that the midpoint of the old interval replaces the point at X8 as one of the end points. The other end is still at X9. The program then goes to line 530 where the new interval is tested.

On the other hand, at 610, the product might be negative, indicating that the root is between the midpoint and X8. The program then goes to 620 where the midpoint replaces X9 as one of the end points. The other end is still at X8. The value of the function at the midpoint is placed in X3, the position of the midpoint is placed in X9, and the program goes to 530 to test the new interval.

At 540 the width of the interval is compared to the desired precision. If the width is smaller than required, the machine displays the value of t at the midpoint of the last interval and stops.

Figure 2-3. Flow chart for the binary search method.

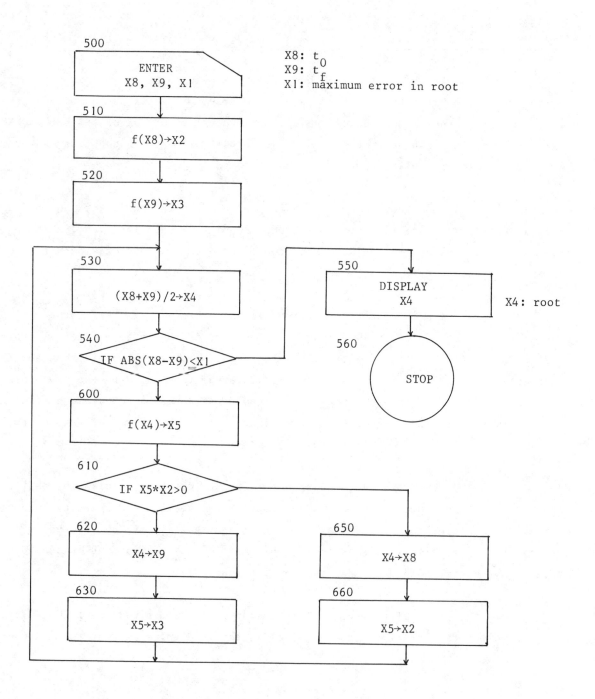

2.4 A Plotting Program

It is usually worthwhile to make a graph of the function before finding the roots. Often a rough estimate of a root can be read from the graph and two values of the independent variable which straddle the root can be picked and then used to start the binary search program. Used in this way, the graph eliminates the need for the uniform interval program. The graph is valuable for other reasons, of course. It gives a clear picture of the function and provides a tool for thinking about the function.

The programmable calculator is extremely useful for making graphs since it can be programmed to perform essentially the same calculation, namely evaluation of the function, over and over again. It thus relieves the tedium of graph making. The program used is very similar to the one given in Fig. 1-1 but there are some differences so we give the entire program here.

Fig. 2-4 shows a flow chart for a plotting program. The independent variable is stored in X1 and the function is a function not only of X1 but also of 2 parameters, stored in X4 and X5 respectively. We may wish to plot the position of a particle as a function of time, for example. X1 is then the time and the 2 parameters are the initial position and velocity of the particle. You may program for more or fewer parameters, of course.

The idea is to retain the same values of the parameters and obtain the value of the function for many different values of the independent variable X1. When this has been done, we want to be able to read in new values of the parameters and plot a new graph. For each set of values of the parameters, the function is evaluated at a series of points, spaced with a uniform interval.

The starting value of the independent variable is entered at line 100 and stored in X1. Also entered are the maximum value of the independent variable, stored in X2; the interval width, stored in X3; and the two parameters. The interval width must be positive.

At line 110, the value of the independent variable is compared to the maximum value desired. The first few times around the loop, the comparison statement is false and the machine proceeds to line 120, where the function is evaluated. Both independent and dependent variables are displayed.

Figure 2-4. Flow chart for a program to evaluate a function for plotting.

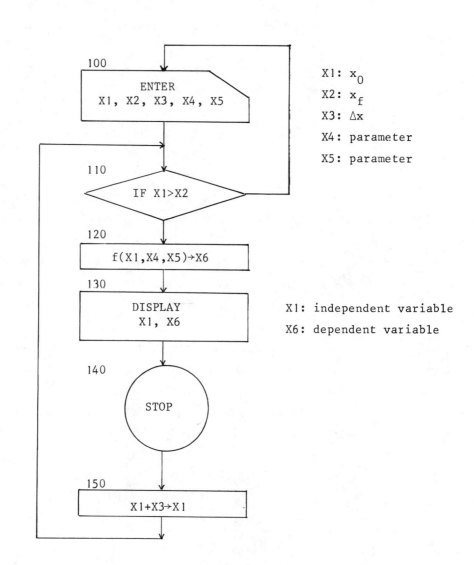

Upon restarting, the machine increments the independent variable by an amount equal to the interval width and goes back to the beginning of the loop, at line 110.

Sooner of later, the value of the independent variable is greater than the maximum desired value. Then the comparison statement at line 110 is true and the machine proceeds to line 100 to accept data for another graph.

2.5 **Exercises**

1. Use the uniform interval method to find intervals which contain the roots of the following functions. Take the interval width to be 0.01 or less.
    a. $x^5 - 7$.
    b. $t^4 + 5.7t^3 - 41.4t^2 - 142.5t + 409.5$
    c. $\sin(x)$ for $-1.1\pi \leq x \leq +1.1\pi$; x is in radians.
    d. $(1/t^3) - \sin(t^3)$ for $-2 \leq t \leq +2$; t^3 is in radians.
    e. $1 - x\ln(x)$ for $0.01 \leq x \leq 5$.

2. For each of the functions of exercise 1, consider the most positive root you found and use the binary search method to find the value of the root with error less than 1×10^{-5}.

3. Consider the function $f(t) = t^3 - 5.4t^2 - 66.3t - 249$.
    a. Make a graph of the function for $-10 \leq t \leq +15$.
    b. This function has one real root. Use the graph to determine an interval which contains the root. Then use the binary search method to find the value of the root with an error less than 1×10^{-5}.
    c. By examination of the graph, choose a new value for the constant term (the last term) of the expression for the function so that the new function has three real roots separated by more than 5. Use the binary search program to evaluate the roots to within 1×10^{-5} of their true values.

4. Find the roots of $f(x) = 15x^3 - 5x^2 + 3$ and evaluate the derivative of f at each root. To do this, modify the binary search program so that, once the root is found, the program evaluates and stores the derivative, then displays the root,

and finally displays the value of the derivative at the root. Find the roots with sufficient accuracy that the error in the derivative is less than 1×10^{-3}.

5. Find the values of t in the range $-10 \leq t \leq +10$ for which the function

$$f(t) = \frac{\sin t}{t}$$

has a local maximum or minimum (its derivative vanishes) and evaluate f(t) for each of the roots of the derivative. The program you wrote for problem 4 can be used. Find the root of the derivative with an error less than 1×10^{-5}.

Chapter 3
MOTION IN ONE DIMENSION

In this chapter, root finding techniques are applied to kinematic problems in one dimension. Use of the calculator allows us to consider situations for which the position of the object is a more complicated function of time than for more standard problems. In addition, the problems of this chapter are intended to help the student to a deeper understanding of the kinematic equations and the information which can be obtained from them. This chapter should be studied in conjunction with Chapter 3 of PHYSICS, Chapter 3 of FUNDAMENTALS OF PHYSICS, or equivalent material in another text. While reading the text book material, the student should try to recognize many of the standard problems which involve the finding of a root.

3.1 Time as the Root of Kinematic Functions

For the following discussion, we assume that the object under consideration is moving along the x axis and that its position x(t) is known as a function of time. For some problems, it is necessary to know, in addition, the x component of its velocity. This may be given or it may be found by differentiating x(t). The symbol v(t) is used to denote the x component of the object's velocity.

In order to have a concrete example in mind as you read the following paragraphs, you may think of the equations for motion with constant acceleration a:

$$x(t) = x_0 + v_0 t + \tfrac{1}{2}at^2 \qquad (3-1)$$

and
$$v(t) = v_0 + at \qquad (3-2)$$

where x_0 is the position at t=0 and v_0 is the x component of velocity at t=0. Remember, however, that for many problems, x(t) and v(t) are much more complicated functions of time.

One of the most basic problems of kinematics is to find the time for which the position of the object is some given number x_A. That is, given x(t) and x_A, for what time t_A is $x(t)=x_A$? The problem can be written in a standard form if we define a new function f(t) by

$$f(t) = x(t) - x_A. \tag{3-3}$$

Then we wish to know the value of t for which f(t)=0. This is precisely the problem for which the programs of the last chapter were devised.

For some problems, the object reaches a maximum x, turns around, and retraces its path. For these problems, there may be two roots in the region searched. There may be more if the object turns around more than once. Care must be taken to ensure you get the roots you are interested in.

Once the root t_A of f(t) has been found, some secondary questions can be answered. These might be: what is the velocity of the object when it reaches x_A? or, what is the acceleration of the object when it reaches x_A? To answer these questions, one simply evaluates v(t) or a(t) at $t=t_A$. If the object reaches x_A more than once, these questions should be phrased more precisely to specify for which time v(t) or a(t) are to be evaluated.

Another basic question, similar to the first, is: at what time does the x component of the velocity have a given value v_A? To answer this question, we define the function

$$f(t) = v(t) - v_A \tag{3-4}$$

and look for its roots. Once the root t_A has been found, it can be used to find the position of the object when its x component of velocity is v_A. Simply evaluate x(t) for $t=t_A$.

For many problems, the time and place for which the object comes temporarily to rest are sought. Then $v_A=0$ and $x(t_A)$ is the maximum or minimum displacement of object from the origin, at least for times near t_A.

Another class of problems for which one seeks the roots of a known function of time has to do with two objects which move along the same line or along parallel lines. Questions asked are: do they collide? (or does one overtake the other?); if they collide, at what time does the collision take place?; at what place does the collision take place?; what are their velocities at the collision?

If $x_1(t)$ is the position of object 1 and $x_2(t)$ is the position of object 2, the collision, if it occurs, takes place at time t_c such the $x_1(t_c)=x_2(t_c)$. Define the function

$$f(t) = x_1(t) - x_2(t). \tag{3-5}$$

Then t_c is the root of $f(t)$. For a collision, the physical solution sought is the smallest root, since the functions which describe the motion are different after the collision. If the objects are on separate but parallel paths, perhaps more than one of the roots is of interest.

Once the root t_c has been found, the other questions can be answered by evaluating one or more of the functions $x_1(t)$, $x_2(t)$, $v_1(t)$, and $v_2(t)$ for $t=t_c$. The velocities can be found, of course, by direct differentiation of the position functions.

3.2 Parameters of the Motion as Roots

For some problems, the time t_A for which an object is at a certain position x_A is given. This information allows us to solve for the value of some quantity, other than the time, which appears in the equation for $x(t)$. For example, suppose it is known that an object moves with constant acceleration along the x axis. Then its position as a function of time is given by Eqn. 3-1. If it is also known that $x(t_A)=x_A$, then we may write $x_0 + v_0 t_A + \frac{1}{2}at_A^2 - x_A = 0$. This equation can be solved for v_0 if x_0, x_A, a, and t_A are known. Furthermore, the root finding programs can be used. We define the function

$$f(v_0) = x_0 + v_0 t_A + \frac{1}{2}at_A^2 - x_A, \tag{3-6}$$

a function of v_0, and seek its roots.

Alternatively, x_0, x_A, v_0, and t_A might be given and we are asked to solve for the acceleration a. We then think of

$$f(a) = x_0 + v_0 t_A + \frac{1}{2}at_A^2 - x_A, \tag{3-7}$$

as a function of a and solve for its root.

The examples described are quite simple and a calculator program is not required for their solution. On the other hand, x(t) might be considerably more complicated and the same questions asked. The only way to answer them might be via the root finding programs.

For more complicated functions, of course, the acceleration is not constant and it is not valid to ask for the acceleration as a parameter. Sometimes, however, the mathematical form of x(t) is known and the equation contains some unknown parameter. Numerical methods can be used to solve for the value of the parameter.

For example, suppose it is known that the position of an object as a function of time has the form

$$x(t) + \frac{v_0}{k}(1 - e^{-kt}) \tag{3-8}$$

where k is a constant and v_0 is the initial velocity. Suppose further that v_0 is known but k is not. The parameter k can be found if it is known that the object is at $x=x_A$ at time $t=t_A$. In fact, k is the root of

$$f(k) = \frac{v_0}{k}(1 - e^{-kt_A}) - x_A. \tag{3-9}$$

In general, x may be a function of time t and some parameter b. We write x(t,b). The information that $x=x_A$ when $t=t_A$ can be used to solve for b. It is the root of

$$f(b) = x(t_A, b) - x_A. \tag{3-10}$$

Similarly, the velocity may be a function of a parameter b and the information that $v=v_A$ when $t=t_A$ can be used to solve for b. It is now the root of

$$f(b) = v(t_A, b) - v_A. \tag{3-11}$$

These general considerations, of course, also apply to the previous discussion. The parameter b might be, in a particular case, the initial velocity v_0 or the constant acceleration a.

3.3 Exercises

For each of the following exercises, set up the appropriate function so that the first step toward the solution consists of finding the root of the function. Then solve for the root numerically using the programs of Chapter 2. If possible, solve the exercise analytically and compare answers.

1. A ball is thrown straight upward in the earth's gravitational field ($g=9.8 m/s^2$). If the initial speed is 20 m/s, how long does it take to reach its maximum height and what is the maximum height? It is thrown from ground level.

2. A ball is thrown straight upward in the earth's gravitational field and is observed to pass the top of a 20 m high building 1.2 s after it is thrown. It is thrown from ground level.
   a. What is its initial velocity?
   b. What is its velocity when it passes the top of the building?
   c. At 1.2 s after the throw, is the ball going up or down?
   d. If it is going up, find the time when it again passes the top of the building. If it is going down, find the time when it first passes the top of the building.

3. The position of a particle moving along the x axis is given by

$$x(t) = 5 + 3.17t^2 - 0.85t^3$$

   where x is in meters and t is in seconds.
   a. At what time is the particle at x=0? Is there more than one time? Substitute your answer into x(t) and evaluate. Did you indeed find a root?
   b. What is the velocity of the particle when it is at x=7.5 m? Is it there more than once?
   c. What is the maximum positive value of x attained by the particle?
   d. Is the x component of the velocity ever -6.3 m/s? If it is, find the time and the position of the particle when it is.

e. Is the x component of the velocity ever +4.5 m/s? If it is, find the time and the position of the particle when it is.

4. When a traffic light turns green, an automobile starts from rest and accelerates with a constant acceleration of 1.8 m/s^2. At the same instant, a truck, travelling with a constant speed of 10.3 m/s overtakes and passes the auto.
    a. At what time does the auto next overtake the truck?
    b. How far beyond the light does the auto overtake the truck?
    c. How fast is the auto travelling when it overtakes the truck?

3.4 Motion in a Uniform Gravitational Field with a Drag Force

In this section we consider an object which moves straight up or down in the earth's gravitational field. It is in air or some other fluid (such as water) and it is acted upon by the earth and by the fluid. For many objects, the force of the fluid is directly proportional to the speed of the object and is directed opposite to the velocity of the object. If we place the y axis along the line of motion and let positive y be in the up direction, then the acceleration of the object can be written

$$a = -g - bv .\qquad (3\text{-}12)$$

The constant of proportionality b depends on the mass of the object and on details of the interaction between the fluid and the object. Factors such as the smoothness and shape of the object's surface as well as the density and viscosity of the fluid enter into a theoretical determination of b.

The constant b also appears in the equation for the position of the object as a function of time. If y=0 for t=0, then

$$y(t) = \frac{1}{b}\left(\frac{g}{b} + v_0\right)(1 - e^{-bt}) - \frac{gt}{b} \qquad (3\text{-}13)$$

where v_0 is the initial velocity (positive if upward, negative if downward).

Problem 1. Differentiate Eqn. 3-13 to find an expression for the velocity as a function of time. In particular, show that

$$v(t) = (\frac{g}{b} + v_0) e^{-bt} - \frac{g}{b}.$$

Show that this expression reduces to $v=v_0$ for $t=0$. Then differentiate the velocity and verify that the acceleration is given by Eqn. 3-12.

We wish to investiage these equations and, by so doing, learn about objects moving in a resistive medium.

Problem 2. Take $b=0.1 \text{ s}^{-1}$ and, on the same paper, plot 3 graphs of y as a function of time, one for each of the following values of v_0: -150 m/s, +25 m/s, and +50 m/s. Consider values of t from 0 to 20 s and let the y axis run from -1000 m to +200 m. For Eqn. 3-13, y=0 represents the starting position of the object (at t=0). Negative values of y are permissible and represent positions of the object below the original position. They occur when, for example, the object is thrown downward from the edge of a cliff.

Notice that all of the curves eventually become straight lines and these lines all have the same slope. After a sufficiently long time, the object moves downward with a constant speed and it is the same constant speed regardless of the initial velocity. The velocity is said to approach the <u>terminal velocity</u>.

The terminal velocity is the velocity for which the acceleration vanishes and, from Eqn. 3-12, this is

$$v_{term} = -g/b. \tag{3-14}$$

For $b=0.1 \text{ s}^{-1}$, the terminal velocity is about 98 m/s.

If the object is thrown downward with an initial speed greater than the terminal speed (the v_0=-150 m/s curve), air resistance slows it until it reaches terminal speed. If the object is thrown downward with an initial speed less than the terminal speed or is thrown upward with any initial speed, the force of gravity accelerates it downward until it is travelling downward with the terminal speed.

By drawing a horizontal line across the graph at some given value of y, it is easy to find the time it takes for the object to reach that position. For example, if the object is thrown upward from the edge of a cliff at 50 m/s and misses the cliff edge on the way down, it hits the canyon floor 400 m below in about 16.2 s. This is determined by drawing the line at y=-400 m and reading the time coordinate of the intersection of this line with the curve for v_0=50 m/s.

<u>Problem 3.</u> Take v_0=+50 m/s and, on the same paper, plot graphs of y as a function of time, one for each of the following values of b: 0 (use $y=v_0 t-\frac{1}{2}gt^2$), 0.01 s^{-1}, and 0.1 s^{-1}. Consider times from 0 to 20 s and let the y axis run from -1000 m to +200 m.

These graphs help us think about y(t) for various values of b. We notice that, as the resistance becomes stronger (b becomes larger), the terminal speed decreases, as we expect from Eqn. 3-14. The terminal speed can be found approximately from the graph by estimating the slope at the far right of the graph. Do this for each curve and compare the value with g/b.

We also notice that, as b increases, the highest point on the curve is reached sooner and that this point is lower than the corresponding point for smaller b.

For objects thrown upward with the same initial speed and allowed to fall below the starting point, it is not necessarily true that the object with the least resistance gets to a given height first. A decrease in b makes the object spend more time going up so it may get to the chosen level later than another object which does not spend as much time on the upward portion of its trajectory.

This phenomenon is shown by the various curves crossing each other. The b=0.1 s^{-1} curve crosses y=-200 m at an earlier time than the b=0.01 s^{-1} curve. The b=0.01 s^{-1} curve crosses y=-600 m at an earlier time than the b=0.1 s^{-1} curve. Of course, after enough time has gone by, all objects will have achieved terminal speed and the one with the smallest b will be going the fastest and will eventually overtake the others.

The following problems deal with one dimensional motion in a resistive medium. For the first three, the position of the object follows one of the graphs you have already plotted. Use the appropriate graph to approximate the solution before using one of the root finding programs to obtain a more accurate solution.

Problem 4. Consider an object which is thrown upward from the ground with an initial speed of 50 m/s. Take $b=0.1$ s^{-1} and $g=9.8$ m/s^2.

a. Find the time it takes to reach the highest point of its trajectory.
b. How high is the highest point?
c. What is its speed when it is halfway to the highest point on the way up?
d. How long after it is thrown does it hit the ground?
e. How fast is it going when it hits the ground? Give your answer in m/s and as a ratio to the terminal speed.

Problem 5. A ball is thrown straight upward from the edge of a cliff with an initial speed of 25 m/s. On the way down, it misses the cliff edge and continues to fall to the canyon floor 350 m below. Take $b=0.1$ s^{-1}.
a. How long is the ball in flight?
b. What is its speed when it hits the canyon floor?
c. What is the ratio of its final speed to its terminal speed?

Save your answers to these questions. They will be used later for an exercise in Chapter 8.

Problem 6. Two objects are simultaneously thrown upward from the edge of a <u>cliff</u>, each with an initial speed of 50 m/s. On the way down, they miss the cliff edge and continue to fall. One object is blunt and, for it, $b=0.1$ s^{-1} while the other is sharp nosed and, for it, $b=0.01$ s^{-1}. Where do they meet?

Problem 7. A ball is thrown upward from the ground with an initial speed of 25 m/s. It reaches the highest point on its trajectory in exactly half the time it would have taken to reach the highest point had there been no air resistance.

a. What is the value of b?
b. How high is the highest point of its trajectory?

Problem 8. Three objects, with b=0.05 s^{-1}, $0.2 s^{-1}$, and 0.4 s^{-1} respectively, are simultaneously shot upward from the edge of a cliff, each with an initial speed of 50 m/s. On the way down, they miss the cliff edge and keep falling. Which reaches the ground 100 m below the cliff edge first? How much later does each of the others arrive?

Problem 9. The curves plotted in problems 2 and 3 can be compared to similar curves for objects moving in a vacuum (zero resistance). Two statements which are true for the objects in a vacuum are:

1. For an object which is thrown upward, the time to reach the highest point and the time to return to the starting height are both proportional to the initial speed.
a. For an object which is thrown upward, the time to return to the starting height is exactly double the time to reach the highest point.

Are similar statements true for an object thrown upward in a resistive medium? If you find from your graphs that they are not, what qualitative statements can you make and justify which might replace the statements made above?

Chapter 4
CURVE FITTING

For much numerical work it is necessary to use an analytic representation of a function, but the function is known only at a series of discrete values of the independent variable. In this chapter we show how a function can be represented, over a small range of the independent variable, by a polynomial and we present a technique for finding the polynomial, given a set of values for the function. This is an important numerical technique, which can be applied to the solution of a wide variety of problems. The material presented here is used, in various contexts, throughout the remainder of the book.

4.1 The Fourth Order Polynomial

Given the function $f(t)$ in the neighborhood of some point t_0, it is possible and often advantageous to approximate it by a simpler function. We discuss in detail one possibility, a fourth order polynomial. The chief reason for wanting to do this is that the polynomial is easy to manipulate mathematically: it can be evaluated, differentiated, and integrated with ease; easier, it is hoped, than the function it replaces. Sometimes the functional form is unknown as, for example, when the function is known only as a series of values produced by a calculator.

We assume that the function $f(t)$ is known in the sense that it can be evaluated at 5 or more points in the neighborhood of $t=t_0$. The points we are interested in are t_0-2h, t_0-h, t_0, t_0+h, and t_0+2h. They are centered on t_0 and the interval between successive points is the constant h.

If $f(t)$ is given in analytic form, then we can evaluate it at any points we choose and, in particular, we may select points with an interval h. If, on the other hand, $f(t)$ is given in tabular form (as from a handbook or a collection of data), then we cannot select h but must take the data as it comes. For the moment, we leave the choice of h open.

To make some of the equations which follow less cumbersome, we define the set of numbers y_{-2}, y_{-1}, y_0, y_1, and y_2 by

$$y_{-2} = f(t_0-2h), \qquad y_{-1} = f(t_0-h), \qquad y_0 = f(t_0),$$

$$y_1 = f(t_0+h), \qquad \text{and} \qquad y_2 = f(t_0+2h), \qquad (4-1)$$

respectively. These are simply the values of the function at the selected points.

It is desired to approximate $f(t_0+\Delta t)$ by a polynomial of the form

$$A_0 + A_1(\Delta t) + A_2(\Delta t)^2 + A_3(\Delta t)^3 + A_4(\Delta t)^4 \qquad (4-2)$$

where A_0, A_1, A_2, A_3, and A_4 are constants. The constants are chosen so that the polynomial exactly reproduces the values of the function at the five selected points. That is, for $\Delta t=-2h$, we want

$$A_0 - 2hA_1 + 4h^2A_2 - 8h^3A_3 + 16h^4A_4 = y_{-2}. \qquad (4-3)$$

This equation is obtained by substituting $\Delta t=-2h$ into Eqn. 4-2 and setting the result equal to y_{-2}, the value of the function at $t=t_0-2h$. Similar equations can be written for $\Delta t= -h$, 0, $+h$, and $+2h$. They are

$$A_0 - hA_1 + h^2A_2 - h^3A_3 + h^4A_4 = y_{-1}, \qquad (4-4)$$

$$A_0 = y_0, \qquad (4-5)$$

$$A_0 + hA_1 + h^2A_2 + h^3A_3 + h^4A_4 = y_1, \qquad (4-6)$$

and

$$A_0 + 2hA_1 + 4h^2A_2 + 8h^3A_3 + 16h^4A_4 = y_2, \qquad (4-7)$$

respectively.

These form 5 simultaneous equations, which can be solved for the 5 unknowns. The results are

$$A_0 = y_0,$$

$$A_1 = \frac{8(y_1 - y_{-1}) - (y_2 - y_{-2})}{12h},$$

$$A_2 = \frac{16(y_1 + y_{-1}) - (y_2 + y_{-2}) - 30y_0}{24h^2},$$

$$A_3 = \frac{(y_2 - y_{-2}) - 2(y_1 - y_{-1})}{12h^3},$$

and

$$A_4 = \frac{(y_2 + y_{-2}) - 4(y_1 + y_{-1}) + 6y_0}{24h^4}. \qquad (4\text{-}8)$$

Given the values of the function at 5 points, equally separated, these equations can be used to find the coefficients for a fourth order polynomial which approximates the function in the neighborhood of t_0.

The polynomial exactly reproduces the value of the function at t_0-2h, t_0-h, t_0, t_0+h, and t_0+2h. If the function does not change drastically between these points, the polynomial also faithfully reproduces it at points between the 5 selected points.

The polynomial may also reproduce the function at points outside the range of the 5 selected points (that is, for $\Delta t < -2h$ or $\Delta t > +2h$) but the further Δt is outside the range, the greater the difference we expect between the function and the polynomial used to approximate it.

Generally, the accuracy of the polynomial can be improved by decreasing the value of h. This is because, for smooth functions, changes in the function are smaller over smaller intervals and, if h is reduced, the polynomial is forced to reproduce the function at points which are closer together. For most of the functions we encounter, we can improve a bad fit by reducing h and fitting over a smaller range. Usually we have the choice of obtaining high accuracy over a small range or less accuracy over a large range.

Being forced to use a smaller range when we reduce h is not a serious disadvantage for most of our applications. If we need to fit over a larger range, we can use more than one polynomial, with each polynomial used to represent the function in some small part of the range.

One reaches a point of diminishing return, however, in the process of reducing h to achieve greater accuracy. This comes about because h can become so small that the numbers calculated for the coefficients lose significance. As h becomes small, y_{-1} and y_1 have more nearly the same values. For example, suppose y_{-1} and y_1 are each calculated accurately to 8 significant figures but h is so small that the first 6 are the same; y_{-1} and y_1 differ only in the last 2 significant figures. When the difference is taken, as it is to find A_1 and A_3, the result is accurate to only 2 significant figures. We want to pick h small enough that the polynomial is a good representation of the function but not so small that the loss of significance is intolerable. For the problems given in this book, this is easy to do but for other problems, the polynomial fit and numerical techniques based on it might not be good choices.

The procedure given for finding the coefficients for a fourth order polynomial can easily be adopted to generate the values of the coefficients for a polynomial of any order. For an nth order polynomial, the highest order term is $A_n(\Delta t)^n$ and there are n+1 coefficients (A_0, A_1, A_2, ... A_n). We require n+1 equations to solve for these coefficients and so we must evaluate the function at n+1 points. We force the polynomial to reproduce the function at the n+1 points. This produces n+1 equations, which we solve for the n+1 coefficients. This procedure will be used in Chapter 8 to find a third order polynomial.

4.2 Improved Efficiency

In order to increase computational efficiency, it is usual to write the polynomial in a slightly different form. It is convenient to define a new variable u by the equation

$$u = \Delta t/h \qquad (4-9)$$

and new coefficients by

$$B_0 = A_0, \quad B_1 = hA_1, \quad B_2 = h^2 A_2, \quad B_3 = h^3 A_3,$$

and
$$B_4 = h^4 A_4. \tag{4-10}$$

Then the polynomial is written

$$f(t_0+\Delta t) = B_0 + B_1 u + B_2 u^2 + B_3 u^3 + B_4 u^4 \tag{4-11}$$

with coefficients given by

$$B_0 = y_0,$$

$$B_1 = \left[8(y_1 - y_{-1}) - (y_2 - y_{-2})\right]/12,$$

$$B_2 = \left[16(y_1 + y_{-1}) - (y_2 + y_{-2}) - 30 y_0\right]/24,$$

$$B_3 = \left[(y_2 - y_{-2}) - 2(y_1 - y_{-1})\right]/12,$$

and
$$B_4 = \left[(y_2 + y_{-2}) - 4(y_1 + y_{-1}) + 6 y_0\right]/24. \tag{4-12}$$

Division by h is now done once, when u is calculated, and the repeated division by h has been eliminated from the calculation of the coefficients.

When the polynomial is evaluated, the number of arithmetic operations can be reduced by collecting terms in a different order and writing

$$f(t_0 + t) = (((B_4 u + B_3)u + B_2)u + B_1)u + B_0. \tag{4-13}$$

This form exactly reproduces the old form, Eqn. 4-11, once the indicated multiplications have been carried out and the parentheses have been removed. The old form requires 4 additions and 10 multiplications for its evaluation while the new form still requires 4 additions but only 4 multiplications. This is a considerable saving in computation time if the polynomial is evaluated over again many times.

For the calculator programs which follow, we use Eqn. 4-13 to evaluate the polynomial and Eqn. 4-12 to find the coefficients.

4.3 Use of the Polynomial for Interpolation and Extrapolation

The polynomial can be used to interpolate between tabulated values of the function. As an example, suppose sin(t) is tabulated at every degree and it is desired to find sin 56.8°. Five entries in the table, at 55°, 56°, 57°, 58°, and 59°, can be used to find the coefficients for the fourth order polynomial through these points. For this fit, the central point is $t_0=57°$ and the interval is h=1°.

We suppose the values of B_0, B_1, B_2, B_3, and B_4 have been found. The polynomial can now be evaluated. First compute u for t=56.8°: $u=(t-t_0)/h=(56.8-57)/1=-0.2$, then use this value in Eqn. 4-13. The result is a close approximation to sin 56.8°.

Notice that t_0 is chosen to be the tabulated point closest to the point for which the polynomial is evaluated. In other words, the 5 tabulated points closest to 56.8° are picked to evaluate the coefficients. We could pick 54°, 55°, 56°, 57°, and 58° but then we would use information about the function at 54° and ignore information at 59°.

As another example, consider the equation for the position of an object subject to a constant force of gravity and air resistance, Eqn. 3-13:

$$y(t) = \frac{1}{b} (\frac{g}{b} + v_0) (1-e^{-bt}) - \frac{gt}{b}.$$

Suppose we seek the time of flight t_f as a function of b for a given initial velocity v_0. We need to solve $y(t_f)=0$ for t_f as a function of b. It is not possible to find an analytic solution. However, given a value for b, it is possible to solve for t_f by using the root finding programs of Chapter 2. It is easy to construct a table of t_f for discrete values of b and the table can be used to find the coefficients for a polynomial fit to the function $t_f(b)$. The polynomial can be used to estimate values of t_f for values of b not tabulated.

If the polynomial is evaluated for t_f outside the range of the points used to calculate the coefficients, the process is called extrapolation. Extrapolation is useful if one knows the function between t_1 and t_2, say, and desires to evaluate it outside the range. One then picks 5 points between t_1 and t_2, calculates the coefficients, and evaluates the polynomial at the chosen point.

The process is exactly the same as for interpolation but it carries a strong warning. The accuracy of the polynomial may decrease rapidly as one goes away from the points used for fitting. For interpolation, the point where the polynomial is evaluated is never more than h/2 from a point where the polynomial exactly reproduces the value of the function. For extrapolation, the point where the polynomial is evaluated is not straddled by points for which the polynomial is exact.

Fig. 4-1 shows the flow chart for a program which finds the values of the coefficients B_0, B_1, B_2, B_3, and B_4, then evaluates the polynomial for a point supplied by the user.

X1 contains the central point for the fit (t_0) and X2 contains the interval width (h). These are entered at line 100. At line 110, t_0-2h is calculated, the function is evaluated for this point, and the value is stored in X9. At 120, 130, 140, and 150, the process is repeated for the points t_0-h, t_0, t_0+h, and t_0+2h, respectively, and the values of the function at these points are stored in X10, X3, X11, and X12, respectively. In each case, two statements are placed in a box to save space in writing the flow chart.

Evaluation of the function may consist simply of entering its value from the keyboard (if it already exists in tabular form) or it may consist of evaluating a number of algebraic expressions. In the former case, these boxes can be replaced by the sequence shown in Fig. 4-2. There is no need to compute the current value of the independent variable.

At lines 160, 170, 180, and 190, the coefficients B_1, B_2, B_3, and B_4 are computed. B_0 is simply the value of the function at t_0. It was previously found at line 130 and stored in X3. B_1 is stored in X4, B_2 in X5, B_3 in X6, and B_4 in X7.

If the coefficients are required for some other purpose, the machine can be stopped at this point and interrogated to find their values. If the program is allowed to continue, it evaluates the polynomial at a value of t supplied to it. At line 200, a value of t is entered and stored in X8. At 210, u is calculated as $\Delta t/h=(t-t_0)/h$ and, at 220, the polynomial is evaluated and the result stored in X9. At 230, the value is displayed and the machine stops to allow the user to copy the value. When it is restarted, it proceeds to line 200 to accept another value of the independent variable.

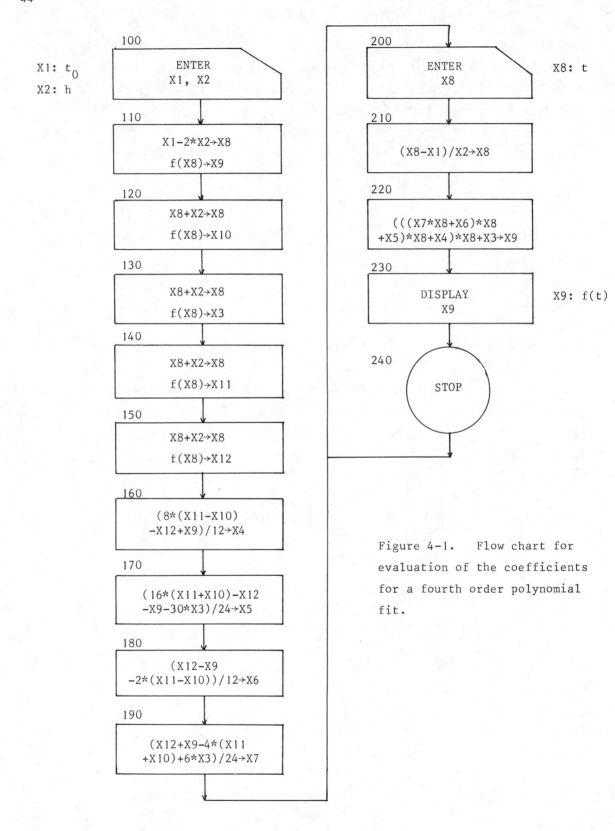

Figure 4-1. Flow chart for evaluation of the coefficients for a fourth order polynomial fit.

Figure 4-2. Modifications to polynomial fit program for use when function is in tabular form.

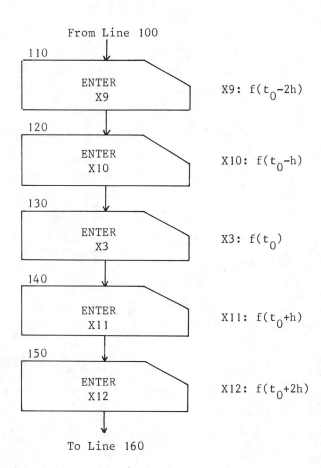

4.4 Use of the Polynomial for Differentiation

The derivative of the fourth order polynomial with respect to the independent variable t is

$$\frac{df}{dt} = \frac{df}{du}\frac{du}{dt} = (B_1 + 2uB_2 + 3u^2B_3 + 4u^3B_4)\frac{1}{h} \qquad (4\text{-}14)$$

where we have used $f(t) = B_0 + uB_1 + u^2B_2 + u^3B_3 + u^4B_4$ and $u=(t-t_0)/h$. Here f is used to denote the polynomial rather than the function it approximates.

Since the polynomial approximates the function in the neighborhood of t_0, we expect the derivative of the polynomial to approximate the derivative of the function in that region. It should be noted, however, that the polynomial which represents the derivative is of a lower order than the polynomial which represents the function and, as a consequence, it is somewhat less accurate. Accuracy can be improved by using a smaller value for the interval h or by using a higher order polynomial approximation. For our work, the third order polynomial approximation to the derivative is sufficiently accurate.

Note that the derivative of the function at $t=t_0$ (u=0) is the coefficient B_1. If we wish to estimate the derivative at some selected point, we usually choose that point as the central point t_0 of the 5 points used for fitting. It is then necessary to calculate only B_1. The other terms in Eqn. 4-14 describe how the derivative changes as one moves away from the central point.

At the minimum or maximum of a function, its derivative vanishes. The value of the independent variable at a minimum or maximum can be found by searching for the root of the expression in Eqn. 4-14. First find the value of u for which this polynomial vanishes, then solve for the independent variable t: $t-t_0+uh$, where t_0 is the central point and h the interval width used to find the coefficients.

Eqn. 4-14 can be put into a form which makes its evaluation more efficient. By appropriate factoring, the equation can be written

$$\frac{df}{dt} = (((4B_4u + 3B_3)u + 2B_2)u + B_1)/h. \qquad (4\text{-}15)$$

This form cuts down on the number of arithmetic operations needed to evaluate df/dt.

The interpolation program, Fig. 4-1, can be modified easily to evaluate the derivative of the polynomial. Between lines 220 and 230, insert the instruction

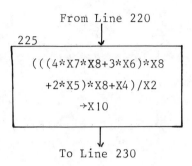

The value of the derivative is stored in X10 and the machine must be asked to display it. Add X10 to the list in the DISPLAY instruction at line 230.

Both the program of Fig. 4-1 and the modification given in Fig. 4-2 can be used to evaluate the derivative.

If the binary search program of Fig. 2-3 is used to find the zero of a polynomial or its derivative, it must be modified somewhat. The coefficients must be entered at the beginning of the program, then used at lines 510, 520, and 600 to evaluate the polynomial or its derivative, as appropriate.

The coefficients B_0, B_1, B_2, B_3, and B_4 are stored in X10, X11, X12, X13, and X14 respectively and, prior to line 500 of Fig. 2-3, the statement

```
       ENTER
X10, X11, X12, X13, X14
```

should be added. If a root of the polynomial is sought, line 510 should read

(((X14*X8+X13)*X8+X12)*X8+X11)*X8+X10→X2

line 520 should read

$$(((X14*X9+X13)*X9+X12)*X9+X11)*X9+X10 \to X3$$

and line 600 should read

$$(((X14*X4+X13)*X4+X12)*X4+X11)*X4+X10 \to X5$$

If a zero of the derivative is sought, line 510 should read

$$((4*X14*X8+3*X13)*X8+2*X12)*X8+X11 \to X2$$

line 520 should read

$$((4*X14*X9+3*X13)*X9+2*X12)*X9+X11 \to X3$$

and line 600 should read

$$((4*X14*X4+3*X13)*X4+2*X12)*X4+X11 \to X5$$

It is not necessary to divide by h since we seek a root.

To use this program, two points which straddle a root must be entered as values of u. That is, if the root is suspected to lie between t_1 and t_2, then $u_1=(t_1-t_0)/h$ is entered into X8 and $u_2=(t_2-t_0)/h$ is entered into X9. Values of t_0 and h must be the same as those used in the fitting program to find the B's.

Similarly, the number stored in X4 and displayed as the root is a value of u. To convert to the value of the independent variable, use $t=t_0+uh$.

4.5 Exercises

The following exercises have been designed to help you obtain an understanding of some of the problems involved in curve fitting. They also give you a chance to write and use the programs discussed in this chapter.

1. a. Find the B coefficients for a fourth order polynomial which represents
        $\sin(t)$ for t between 0 and $\pi/2$ radians. Take $t_0=\pi/4$ and $h=\pi/8$.
    b. Find the fractional error (the difference between the polynomial and the
        exact function divided by the exact function) for values of t taken every
        $\pi/10$ between 0 and $\pi/2$ radians.
    c. The derivative of this function is $\cos(t)$, another function which is easily
        evaluated on the calculator. Evaluate the derivative of the polynomial and
        the fractional error for values of t taken every $\pi/10$ from 0 to $\pi/2$ radians.

2. a. Use the same function as in exercise 1 but take $h=0.05$ radians (retain
        $t_0=\pi/4$ radians) and evaluate the B coefficients.
    b. Find the fractional error in the polynomial for values of t taken every
        0.07 radians from $\pi/4-0.22$ to $\pi/4+0.06$ radians. For the first two
        points, the polynomial is used to extrapolate while for the rest it is
        used to interpolate.
    c. Evaluate the fractional error in the derivative of the polynomial. Use
        the same values of t as in part b.

3. a. Let $f(t)=\exp(-t^2)$ and evaluate the B coefficients for a fourth order
        polynomial approximation to this function. Take $t_0=0$ and $h=0.5$.
    b. Compare the function and the polynomial at $t=0, 0.2, 0.4, 0.6,$ and 0.8.
        Do this by computing the fractional error.
    c. Now fit the same function about $t_0=0.1$ with $h=0.5$ as before. Evaluate the
        B coefficients and calculate the fractional error at $t=0, 0.2, 0.4, 0.6,$
        and 0.8.

4. a. Fit the function $f(t)=(\sin t)(\cos t)$ to a fourth order polynomial valid
        in the neighborhood of $t=40°$. Use $t_0=40°$ and $h=3°$.
    b. Use the binary search program to find the root of the derivative of
        $f(t)$. That is, find the value of u for which $df(t)/dt$ vanishes, then
        compute the value of t corresponding to that value of u.
    c. The root of df/dt can easily be found by analytic means. We note that
        $(\sin t)(\cos t)=\frac{1}{2}\sin 2t$ and this has a maximum when $2t=90°$ or $t=45°$.
        Calculate the fractional error in the value of t at the root as found by
        the polynomial method.

5. In section 3.4, we considered several objects which are thrown straight up with the same initial speed but which are subjected to different drag forces. It was found that, if they are allowed to fall to an end point below the starting point, an object subject to a larger drag force might reach the end point sooner than an object subject to a smaller drag force. It was conjectured that there exists a value for the drag coefficient b such that the time to reach the end point is a minimum. We are now in a position to calculate that value of b.

   We consider objects which are thrown upward with an initial speed of 50 m/s from the edge of a cliff. On the way down, they miss the cliff edge and continue to fall to the ground 200 m below. For what value of b does the trip take the least amount of time and what is the time of flight for that value of b?

   Use Eqn. 3-13 and one of the root finding programs to calculate the time of flight for various values of b. Start with b=0 (use $y=v_0 t-\frac{1}{2}gt^2$ for this case) and increment by 0.01 s^{-1}. Stop when you have 2 values for the time of flight beyond the minimum.

   Make a plot of t_f vs. b to be sure the function is reasonably smooth and, if you are satisfied, fit the function to a polynomial, using the point for which the time of flight is smallest for the central point. You also need 2 points on each side. Finally, use a root finding program to find the root of the polynomial's derivative.

   To find the smallest time of flight, substitute the value you found for b into Eqn. 3-13 and use a root finding program to find the time for which y=-200 m.

6. An object is thrown straight up with an initial speed of 15 m/s. The highest point on its trajectory is exactly half the height it would be if there were no drag force. Find the value of the drag coefficient b.

   Let h(b) be the maximum height as a function of b. Show that $h(0)=\frac{1}{2}v_0^2/g$. Find h(b) for various values of b and fit the function $f(b)=h(b)-\frac{1}{4}v_0^2/g$ to a polynomial using points on either side of f(b)=0. Find the root of f(b). Start with b=0.8 s^{-1} and, for each trial, increment it by 0.1 s^{-1}. Stop at 1.4 s^{-1}.

Chapter 5
MOTION IN TWO DIMENSIONS

In this chapter, root finding and polynomial fitting techniques are applied to projectile and uniform circular motion. Drag forces on a projectile are considered. The material in this chapter is intended to supplement Chapter 4 of PHYSICS, Chapter 4 of FUNDAMENTALS OF PHYSICS, or similar material in other texts.

5.1 Projectile Motion

In order to define the symbols used in this section, a typical projectile trajectory is shown in Fig. 5-1. The projectile is fired from the point x_0, y_0 with initial velocity $\vec{v}_0 = v_{0x}\hat{i} + v_{0y}\hat{j} = v_0(\cos\theta\hat{i} + \sin\theta\hat{j})$ where v_0 is the initial speed and θ is the firing angle, measured from the horizontal. The projectile lands at the place with coordinates x_f, y_f. We have chosen the y axis to be parallel to the direction of the acceleration due to gravity and positive in the up direction. The acceleration due to gravity is $-g\hat{j}$ with $g=9.8$ m/s^2. The x axis is chosen so that the trajectory lies in the x,y plane and so that v_{0x} is positive. The quantity $R = x_f - x_0$ is called the range of the projectile.

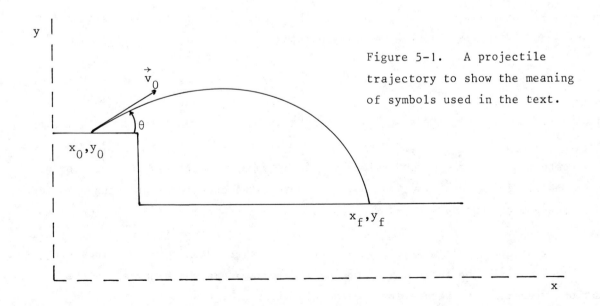

Figure 5-1. A projectile trajectory to show the meaning of symbols used in the text.

The motion of the projectile is described by giving its coordinates x and y as functions of time. For example, if no forces other than that due to gravity act, then

$$x(t) = x_o + v_{0x} t \tag{5-1}$$

and
$$y(t) = y_0 + v_{0y} t - \tfrac{1}{2} g t^2 . \tag{5-2}$$

These equations may be differentiated with respect to time to obtain the components of the velocity as functions of time:

$$v_x(t) = v_{0x} \tag{5-3}$$
and
$$v_y = v_{0y} - gt . \tag{5-4}$$

The projectile may also be subjected to air resistance. If so, there is an additional term in the acceleration; we take it to be proportional in magnitude to the speed of the projectile and directed opposite to the velocity of the projectile. That is,

$$\vec{a} = -g\hat{j} - b\vec{v}(t) . \tag{5-5}$$

Then,
$$x(t) = x_0 + \frac{v_{ox}}{b} (1 - e^{-bt}) \tag{5-6}$$

and
$$y(t) = y_0 + \frac{1}{b} \left(\frac{g}{b} + v_{0y}\right) (1 - e^{-bt}) - \frac{gt}{b} . \tag{5-7}$$

Here b is the constant of proportionality. These equations may be differentiated with respect to time to find the velocity components:

$$v_x(t) = v_{0x} e^{-bt} \tag{5-8}$$

and
$$v_y(t) = \left(\frac{g}{b} + v_{0y}\right) e^{-bt} - \frac{g}{b} . \tag{5-9}$$

The y components of position and velocity are the same respectively as those for one dimensional motion given in section 3.4. To obtain the equations for the x components, simply set g=0 and change y to x.

In general, we assume $x(t)$, $y(t)$, $v_x(t)$, and $v_y(t)$ are known functions of time. They may also be functions of one or more parameters which appear in the equations. Examples of parameters are the coordinates of the initial position, components of the initial velocity, the initial speed, the firing angle, and the drag coefficient.

Many projectile problems can be formulated in terms of finding a root of one of the four basic functions (x, y, v_x, v_y) or closely related functions. For example, at the highest point of the trajectory $v_y=0$ and this equation can be solved for the time at which the projectile is at the highest point, or, if that time is given, it can be solved for one of the parameters. The value of the time or the parameter is then substituted into the other three equations to find the coordinates of the highest point and the x component of the velocity at the highest point.

Similarly, when the projectile hits the ground, $x(t)=x_f$ and $y(t)=y_f$. One of these is solved for the time of landing and this value substituted into the other three equations to find the impact velocity and the other coordinate of the impact point. Usually y_f is given and x_f is calculated.

In some cases, there exists a relationship between x_f and y_f, brought about by the geometry of the terrain in the neighborhood of the impact point. For example, for the situation depicted in Fig. 5-2, x_f and y_f are related by

$$(y_f-y_0) = (x_f-x_0-d)\tan\alpha \tag{5-10}$$

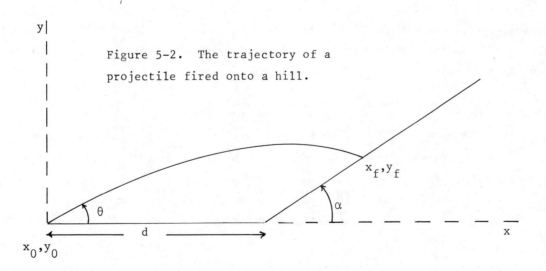

Figure 5-2. The trajectory of a projectile fired onto a hill.

and the time of flight is the root of

$$\left[y(t)-y_0\right] - \left[x(t)-x_0-d\right]\tan\alpha. \qquad (5-11)$$

For other terrains, x_f and y_f are related differently.

If the time of flight is given, these equations may be solved for one of the parameters and then that value and the value of the time may be substituted into the other equations to find information about the projectile at the time of impact.

For all of the problems discussed above, the most important numerical step is to find the root of one of the equations, then substitute the value of the root into the other equations to find the information sought. These problems are very nearly the same as the problems for one dimensional kinematics and the same techniques are used.

Since dealing with one dimensional motion, we have added the capability of representing a function by a polynomial. Among other uses of the polynomial, we learned how to find the maximum or minimum of a function by seeking a root of the derivative of the polynomial which approximates the function. This capability allows us to solve another class of interesting problems: problems which ask for the maximum or minimum value of position or velocity components as functions of one of the parameters.

As an example, consider again the situation shown in Fig. 5-2. Suppose the initial speed v_0 of the projectile is given and we wish to solve for the firing angle for which x_f is the greatest possible value. The steps which lead to a numerical solution are

1. Pick a value for θ.
2. Find the root of $f(t) = \left[y(t)-y_0\right] - \left[x(t)-x_0-d\right]\tan\alpha$ after substituting the functional forms of $x(t)$ and $y(t)$.
3. For that value of t, calculate $R=x(t)-x_0$.
4. Repeat steps 1 through 3 with different values of θ. Five values are needed in all, centered on the one for which R is the greatest.

5. Fit R(θ) to a polynomial.
6. Take the derivative of the polynomial and find its root.

A similar procedure was followed to solve exercise 5 at the end of Chapter 4. This procedure can be carried out for a projectile subject to a drag force as well as for one moving only under the influence of gravity. Only the functional forms of x(t) and y(t) used at step 2 are different.

5.2 Projectile Motion Exercises

1. A projectile is fired over level ground at 150 m/s and at a firing angle of 30° above the horizontal. Neglect air resistance and use root finding programs to calculate:
   a. the time it takes to get to maximum height.
   b. the coordinates of the point of maximum height.
   c. the velocity at maximum height.
   d. the time of flight.
   e. the range.
   f. the velocity at impact.

All of these quantities can be found analytically. Evaluate the analytic expressions and compare answers with those found by numerical techniques. If there are discrepancies, find their origin.

2. This exercise is the same as exercise 1 except that the projectile is subjected to air resistance. A projectile is fired over level ground at 150 m/s and at a firing angle of 30° above the horizontal. It is subjected to air resistance with the force proportional to the velocity. Take $b=0.1$ s^{-1} and find:
   a. the time it takes to get to maximum height.
   b. the coordinates of the point of maximum height.
   c. the velocity at maximum height.
   d. the time of flight.
   e. the range.
   f. the velocity at impact.

3. A projectile is fired from the edge of a cliff onto a plain a distance h below. The initial speed is v_0 and the initial velocity makes the angle θ with the horizontal. Neglect air resistance.

a. Show that the range is given by

$$R = \frac{v_0^2 \sin\theta \cos\theta}{g} + \frac{v_0 \cos\theta (v_0^2 \sin^2\theta + 2gh)^{\frac{1}{2}}}{g}$$

b. Take v_0=150 m/s, h=200 m, and use numerical techniques to find the value of θ for which R is a maximum. What is the maximum value of R?

4. A projectile is fired from the edge of a cliff onto a plain a distance 200 m below. The initial speed is 150 m/s and b=0.2 s^{-1} is the coefficient of air resistance. What is the maximum range that can be achieved by altering the firing angle? What firing angle should be used to achieve the maximum range?

5. A projectile is fired at 60 m/s onto a hill as shown in Fig. 5-2. The base of the hill is 100 m from the firing point and the hill makes an angle of 9.3° with the horizontal. Take the firing angle to be 30° and solve for the following quantities. Obtain 3 significant figure accuracy.
   a. The time of impact.
   b. The position of the impact point.
   c. The distance up the hill to the impact point.
   d. The components of the velocity on impact.

6. Consider the projectile and hill of exercise 5. Find the firing angle which gives the largest x_f. Try firing angles of 40°, 45°, 50°, 55°, and 60° and fit the results to a polynomial. Obtain 3 figure accuracy.

7. A certain hill can be approximated by a hemisphere of radius R. A ball is thrown from the top with initial speed v_0 and at an angle θ to the horizontal. The acceleration due to gravity is downward. The ball lands at the point A such that the radius vector through A makes the angle α with the vertical. Neglect air resistance. Take the origin at the ball's initial position with y positive in the up direction.

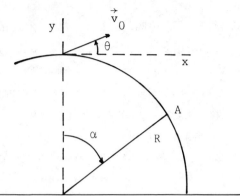

a. Show that the coordinates of the landing place are given by $x = R\sin\alpha$ and $y = -R(1-\cos\alpha)$.
b. Substitute the functional form for $x(t)$ and $y(t)$, eliminate t, and show that the trajectory of the ball intersects the sphere at A if

$$-(1-\cos\alpha)\cos^2\theta = \sin\theta\cos\theta\sin\alpha - \frac{Rg}{2v_0^2}\sin^2\alpha.$$

c. Take R=500 m, v_0=20 m/s, θ=25°, and use a root finding program to calculate α.
d. Take R=500 m, v_0=20 m/s, and find the value of θ for which α is a maximum.

5.3 Uniform Circular Motion

If an object moves with constant speed on a circular path, it is accelerating by virtue of the change in the direction of its velocity. The acceleration is directed radially inward and has the magnitude

$$a = v^2/R \qquad (5\text{-}12)$$

where v is the speed of the object and R is the radius of its path.

In order for the object to accelerate, there must be a force acting on it. In this case the force must be radially inward and must have magnitude given by

$$F = mv^2/R \qquad (5\text{-}13)$$

where m is the mass of the object.

Many forces which occur in nature depend on the position of the object on which they act, measured relative to the positions of other objects. If this type force (position dependent) is used to hold the object in uniform circular motion at a given speed, the radius of the path is determined by the speed and the force law. If an attempt is made to start the object with a speed and radius which are not compatible for circular motion, the force will be either too large or too small to produce the proper acceleration and the orbit will not be circular. We investigate non-circular orbits in a later chapter. For now, we shall be content with finding the radius of the circular path when the object is pulled toward a central point (the origin) by a force which is a known function of position.

As a simple example, suppose a massive spherical object (a sun) is at the origin and a planet is a distance R away. The sun pulls inward on the planet with a force which is inversely proportional to the square of the distance between the sun and the planet. It is also proportional to the product of the masses of the sun and the planet. The constant of proportionality is given the symbol G and has the value 6.67×10^{-11} $m^3/s^2 \cdot kg$. All this is summed up by the force law equation

$$F(R) = G \frac{Mm}{R^2} \qquad (5-14)$$

where M is the mass of the sun and m is the mass of the planet.

If the planet moves on a circle with constant speed, it must be that

$$G \frac{Mm}{R^2} = \frac{mv^2}{R} \qquad (5-15)$$

or, solving for R,

$$R = \frac{GM}{v^2} . \qquad (5-16)$$

There is exactly one radius for which uniform circular motion occurs (for a given speed).

In general, the radius of the orbit is the root of

$$F(R) - mv^2/R \qquad (5-17)$$

where F(R) is the magnitude of the force when the object is distant R from the force center. The force must be radially inward and must be constant in magnitude as the object traverses the orbit.

The following problems give examples of forces which can be used to hold an object in uniform circular motion. In each case use a root finding program to determine the radius of the orbit.

Problem 1. A spring with one end fixed exerts a force on an object attached to the other end if the spring is elongated or compressed. For a so-called ideal spring, the magnitude of the force is given by $k(\ell-\ell_0)$ where k is a constant (called the spring constant), ℓ is the length of the spring, and ℓ_0 is the equilibrium length (when $\ell=\ell_0$ the force is zero). The force is directed along the spring, outward if the spring is compressed and inward if the spring is elongated.

One end of a spring with spring constant k=5 N/m and equilibrium length ℓ_0=0.5 m is fixed to a pivot. The other end is attached to a 0.3 kg mass which moves in a circle with radius equal to the spring length, centered at the pivot. If the speed of the mass is 1.6 m/s, what is the radius of the circle?

Solve this problem using a root finding program. Then solve is analytically and compare answers.

Problem 2. The magnitude of the force for a certain non-ideal spring is given by

$$F = k(\ell-\ell_0) + A(\ell-\ell_0)^2$$

where A is a constant. Take A=1.3 N/m^2, k=5 N/m, and ℓ_0=0.5m. The spring is used to hold a 0.3 kg mass in uniform circular motion at a speed of 1.6 m/s. The radius of the circle is the same as the length of the spring. What is its value?

Problem 3. A 0.025 kg pellet is subjected to a force given by

$$F = 5(0.012 - r^3 e^{-r})$$

where r is the distance from the origin to the pellet and the force is directed inward

when F is positive. F is in newtons and r is in meters.
a. Find the distance from the origin to a place where the pellet remains at rest if placed there. There are two such distances.
b. Suppose the pellet now executes uniform circular motion with speed 1.3 m/s. Find the radius of the orbit.

Chapter 6
INTEGRATION OF NEWTON'S SECOND LAW
WITH APPLICATION TO MOTION IN ONE DIMENSION

In this chapter we present two programs which are used to integrate Newton's second law. The first is quite general in that the force can depend on time, the position of the particle, and the velocity of the particle. The second, which runs faster, can be used only if the force does not depend on velocity. The material in this chapter supplements and enriches Chapters 5 and 6 of PHYSICS, Chapters 5 and 6 of FUNDAMENTALS OF PHYSICS, or similar material in other texts. It provides an extension of the usual introductory discussion to include examples of complicated physical situations. Nevertheless, the only physics involved is the second law itself and constant acceleration kinematics.

6.1 The Basic Program

We suppose a particle of mass m moves along the x axis and is subjected to a force, also directed along the x axis. The x component of the force is denoted by F. Then Newton's second law can be written

$$F = m \, dv/dt, \qquad (6\text{-}1)$$

where $v = dx/dt$ is the x component of the particle's velocity (the only non-vanishing component).

The problem we address is: given the position x_0 and velocity v_0 of the particle at time t_0, what is its position and velocity at some other time t_f (earlier or later than t_0)?

To solve this problem, it is necessary to calculate the force and we suppose an equation is given which allows us to do just that once the time, the position of the particle, and the velocity of the particle are specified.

We give a program to integrate Eqn. 6-1. The programming strategy is to divide the time range from t_0 to t_f into a large number of small intervals, each of width Δt. We assume the width can be adjusted so that the force does not change appreciably from the beginning to the end of any interval and we accordingly take the force to be constant within each interval. The value used for the force may be different for different intervals to allow for changes in the force as time goes on.

With the assumption of a constant force for each interval, Newton's second law can easily be integrated. Let x_{j0} and x_{jf} be the position of the particle at the beginning and end, respectively, of interval j and let v_{j0} and v_{jf} be the velocity of the particle at the beginning and end, respectively, of interval j. Then

$$v_{jf} = v_{j0} + (F_j/m)\Delta t \qquad (6-2)$$

and
$$x_{jf} = x_{j0} + v_{j0}\Delta t + \tfrac{1}{2}(F_j/m)(\Delta t)^2 \qquad (6-3)$$

where F_j is the value of the force used for interval j. These are just the kinematic equations for constant acceleration. The acceleration in interval j is F_j/m.

For each interval, the position and velocity at the beginning of the interval are known, either as a result of the calculation for the previous interval or, in the case of the first interval, from the given initial conditions. Eqns. 6-2 and 6-3 are used to calculate the position and velocity at the end of the interval. These numbers are then used as the position and velocity for the beginning of the next interval and so it goes until t_f is reached.

The value of F_j can be the value of the force function at the beginning of interval j. This is the simplest choice when the force is a function of position and velocity. These quantities are known for the beginning of the interval and their values can easily be substituted into the force law equation. We follow this procedure here and leave the discussion of another choice to the next section.

So far we have been discussing a sequence of steps to calculate the position and velocity of a particle at a specific time t_f, given the force function and the initial values of the position and velocity. Usually one wishes the machine to display the position and sometimes the velocity for a sequence of times so that these quantities can be plotted. On the other hand, one does not usually want the values at the end of every interval. It is often important for accuracy to use a much more narrow interval for calculational purposes than one needs for constructing a graph. The program we present displays the position and velocity only after N intervals have been considered. N is a number supplied by the user. The program then continues for another N intervals before displaying results. At each stage the number of intervals is entered and it can be changed at any stage of the calculation.

The flow chart for the program is shown in Fig. 6-1. Initial data is entered at line 100. Here the quantities stored in the various registers are:

	X1	initial x coordinate (x_0).
	X2	initial x component of velocity (v_0),
and	X3	initial time (t_0).

Immediately, at line 110, the width of an interval is entered and stored in X4 and the number of intervals is entered and stored in X5. The program executes the loop which begins at line 130 exactly X5 times, once for each interval. The position and velocity calculated for the end of the X5 intervals is then displayed. The user must be sure the X5 intervals correspond to the times for which results are desired.

X7 is a counter which keeps track of the number of times the machine starts executing the instructions in the loop. It is zeroed at line 120, outside the loop. At line 130 the loop is entered and the counter is incremented by 1. At the start of the loop, X3 contains the time, X1 contains the position, and X2 contains the velocity for the beginning of the interval being considered. The first time around, these storage locations contain the initial values. At the end of the loop X3 contains the time, X1 contains the position, and X2 contains the velocity for the end of the interval. These are then the initial values for the next execution of the loop instructions.

The acceleration is evaluated at line 150. The function f in this instruction is the force divided by the mass of the particle and it is evaluated using the values of the time (X3), position (X1), and velocity (X2) for the beginning of the interval. The user must supply the specific equation which gives the force in terms of these quantities.

At line 160 the acceleration is multiplied by the interval width and the result is stored in X6. X6 now contains $(F_j/m)\Delta t$. This combination occurs in both the equation for the position and the equation for the velocity and it is more efficient to calculate it once and store it than to calculate it twice. The position at the end of

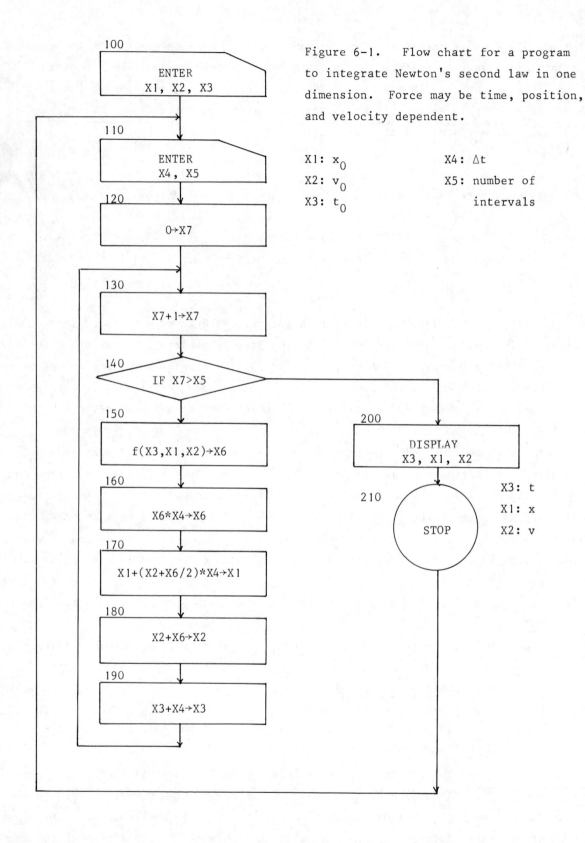

Figure 6-1. Flow chart for a program to integrate Newton's second law in one dimension. Force may be time, position, and velocity dependent.

X1: x_0 X4: Δt
X2: v_0 X5: number of
X3: t_0 intervals

X3: t
X1: x
X2: v

the interval is calculated at line 170 and the velocity at the end of the interval is calculated at line 180. Notice that the position is calculated before the velocity. This is because the equation for the position contains the velocity and we must use the value for the beginning of the interval rather than the value for the end.

The time is updated at line 190. The program then returns to the beginning of the loop to consider the next interval.

The program leaves the loop at line 140 when X7>X5. It has just then finished calculation for the last of the X5 intervals it was asked to consider. At line 200, the position and velocity are displayed and the program stops, at line 210, to allow the user to copy the results.

If the machine is restarted, it proceeds to line 110 where a new interval width and number of intervals is entered. The machine retains the time, position, and velocity for the end of the last interval and these become the initial values for the next sequence of intervals. Sometimes it is desirable to change the interval width so that it is narrow where the force changes rapidly and wide where the force changes more slowly. If no change in Δt or N is anticipated, it is worthwhile to program the machine so that it returns to line 120 rather than 110. This saves entering new data.

6.2 Velocity Independent Forces

If the force does not depend on the velocity of the particle, the program can be modified to run faster and, in many cases, to be more accurate. This can be done because, for these forces, it is just as efficient to calculate the velocity at a different sequence of times than that for which the position and force are calculated.

Recall that
$$x_{jf} = x_{j0} + (v_{j0} + \frac{F_j}{m}\frac{\Delta t}{2})\Delta t \qquad (6-4)$$

is the position at the end of interval j. We note that the quantity between the parentheses is the velocity at the midpoint of the interval (Δt is divided by 2). Let $v_{j\frac{1}{2}}$ be the velocity at the midpoint of interval j. Then

$$x_{jf} = x_{j0} + v_{j\frac{1}{2}}\Delta t. \qquad (6-5)$$

We need to know the velocity only at the midpoints and not the ends of the intervals. The velocity at the midpoint of interval j can be found from the velocity at the midpoint of interval j-1 by using

$$v_{j\frac{1}{2}} = v_{j-1\frac{1}{2}} + (F_j/m)\Delta t \tag{6-6}$$

where F_j is the force at the beginning of interval j.

The equations we use for velocity independent forces are 6-5 and 6-6. There is a certain symmetry about these equations. Fig. 6-2 shows the points in time for which the various quantities are calculated. The value of the force which is used to calculate the velocity for one time from the velocity for a previous time is the value halfway between. Similarly, the value of the velocity which is used to calculate the position for one time from the position for a previous time is the value halfway between.

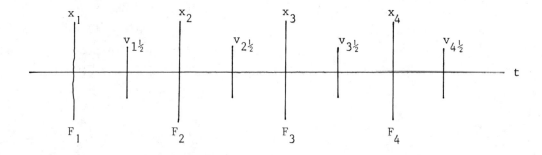

Figure 6-2. Diagram showing the times for which various quantities are calculated when Eqns. 6-5 and 6-6 are used.

The program to be presented is more accurate than the previous one because it uses the value of the force halfway between the times for which the velocities are calculated. For example, F_3 (on the diagram of Fig. 6-2) is used to calculate $v_{3\frac{1}{2}}$, given $v_{2\frac{1}{2}}$. The exact expression for $v_{3\frac{1}{2}}$ is $v_{3\frac{1}{2}} = v_{2\frac{1}{2}} + (F_{ave}/m)\Delta t$ where F_{ave} is the average force in the interval which runs from the point marked $v_{2\frac{1}{2}}$ to the point marked $v_{3\frac{1}{2}}$. Often the value of the force at the midpoint is closer to the average value than the force at the end of an interval.

When writing the calculator program, all intervals can be treated the same except the first and the last. For the first interval, we cannot immediately calculate the velocity using the velocity at the midpoint of the previous interval since we do not know that number. Actually, we can set up the program so that the first interval is the same as the others. We imagine an interval before $t=t_0$ (interval 0) and first calculate the velocity at its midpoint using

$$v_{0\frac{1}{2}} = v_0 - (F_1/m)(\Delta t/2) \tag{6-7}$$

where F_1 is the force at $t=t_0$. The velocity at the midpoint of the first interval is then $v_{1\frac{1}{2}} = v_{0\frac{1}{2}} + (F_1/m)\Delta t$ and Eqn. 6-6 can be used for this interval.

The last interval is different from the others because one usually wants to find the velocity at the end rather than at the midpoint of the interval. When the machine finishes with the last instruction of the loop, it has then calculated the velocity for the midpoint of the last interval. We can use

$$v_f = v_{N\frac{1}{2}} + (F_{N+1}/m)(\Delta t/2) \tag{6-8}$$

to find the velocity at the end point. To be consistent F is taken to be the force at the end of the last interval. If the intervals are sufficiently narrow, this is not an important consideration, however.

Fig. 6-3 gives the flow chart for a program which can be used when the force is independent of the velocity. The program is very similar to the previous one, shown in Fig. 6-1. The meanings of the symbols and the logical pattern are the same.

At line 115, the acceleration at $t=t_0$ is calculated from the force law and, at line 116, this value is used to find the velocity at the midpoint of the interval just prior to t_0. The function f is the force divided by the mass, evaluated for the time stored in X3 (t_0) and the position stored in X1 (x_0).

At line 150, in the loop, the acceleration at the beginning of the interval is calculated, at 170 the velocity at the midpoint of the interval is calculated, and at 180 the position at the end of the interval is calculated. Note that the velocity must now be calculated before the position so that the program uses the value of the

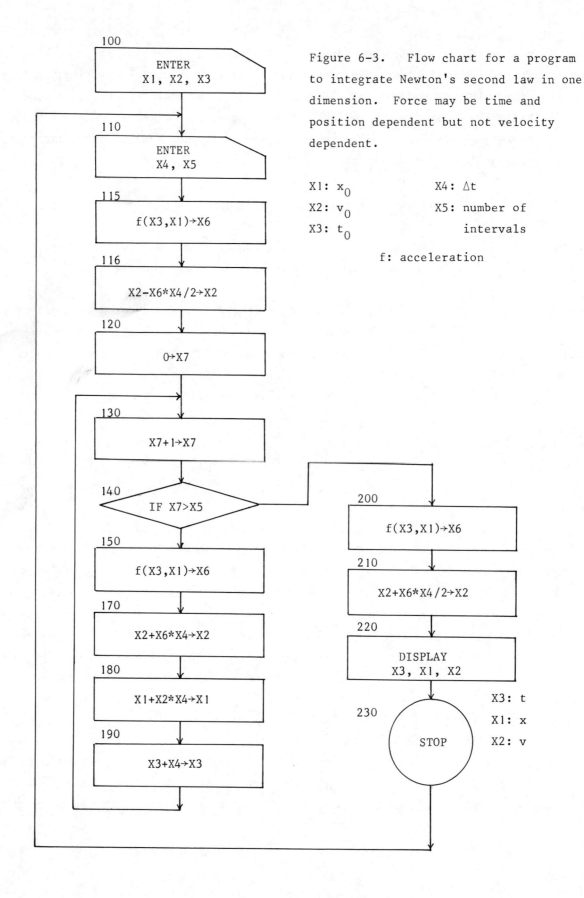

Figure 6-3. Flow chart for a program to integrate Newton's second law in one dimension. Force may be time and position dependent but not velocity dependent.

X1: x_0 X4: Δt
X2: v_0 X5: number of
X3: t_0 intervals

f: acceleration

X3: t
X1: x
X2: v

velocity at the midpoint of the current interval to calculate the position.

When the program exits the loop, at line 200, X1 contains the position at $t=t_f$ but X2 contains the velocity at $t=t_f-\Delta t/2$ (the midpoint of the last interval). At line 200, the force at $t=t_f$ is calculated and used, at line 210, to find the velocity at $t=t_f$. This is stored in X2.

6.3 <u>Tests of Accuracy</u>

For several types of force functions, analytic techniques can be used to find exact solutions to the corresponding Newton's second law equation. By comparison of the answer produced numerically with the exact solution, we can gain some idea of the accuracy of the calculator program.

In general, numerical accuracy can be increased by reducing the time interval Δt over which the force is assumed constant. This, however, increases computation time and one usually picks Δt as large as possible but not so large that the error produced is intolerable. For fixed Δt, the error depends on the rate at which the force is changing and on the number of significant figures carried by the machine. We can obtain some idea of how small the interval should be by trying several interval sizes and comparing answers with the exact answers.

Errors are cumulative. For each interval, values of the position and velocity for the end depend on the values used for the beginning. Any error produced during the calculation for a given interval may add to the error inherent in the beginning values to produce even larger error in the values for the end of the interval. Errors arise from the erroneous assumption that the force is constant in the interval and from the fact that the machine carries only a finite number of significant figures. It is as if the digit to the right of the last digit carried is a zero and this is most likely in error. Because of the cumulative nature of the errors, we expect larger errors in the position and velocity values calculated for later times than in those for earlier times.

If the force is exactly constant, we expect both programs to yield highly accurate results. Try the following problem.

Problem 1.   A 20 kg chest slides across the floor, subject to a 5 N force of friction. Take the initial speed to be 3 m/s and find the position and velocity at the end of 3 s, 6 s, 9 s, and 12 s. The force is directed opposite to the velocity.
a.   Solve the problem exactly, using analytic means.
b.   Use the program of Fig. 6-1 and an interval width of 1 s.
c.   Use the program of Fig. 6-3 and an interval width of 1 s.

If the force is proportional to the time, then the average force in an interval is exactly the same as the force at the midpoint of the interval and we expect the program of Fig. 6-3 to yield results for the velocity which are exact except for round-off error. On the other hand, velocity values produced by the program of Fig. 6-1 are expected to be in error since the force is not constant. Values of the position are in error for both programs but, since the velocity calculated by the program of Fig. 6-3 is more accurate than that calculated by the program of Fig. 6-1, we expect the position as calculated by that program to be more accurate also. The following problem allows us to check these statements.

Problem 2.   A 20 kg chest slides across the floor, subject to a force given by

$$F(t) = 5 - 0.5t.$$

F is in newtons and t is in seconds. Take the initial velocity to be 3 m/s in the direction opposite to that of the force at t=0 and find the position and velocity at the end of 3 s, 6 s, 9 s, and 12 s. Use the following methods.
a.   Analytical integration to find exact values.
b.   The program of Fig. 6-1 with an interval width of 1 s.
c.   The program of Fig. 6-1 with an interval width of 0.1 s.
d.   The program of Fig. 6-3 with an interval width of 1 s.

Notice that the velocities calculated in part d are accurate to nearly the number of significant figures carried by the machine. The positions calculated in part d are much more accurate than those calculated in part b and about as accurate as those calculated in part c. The accuracy increases from part b to part c, but at a price. You were probably aware of the increase in computing time. In the case of part c, the interval is a tenth as wide but there are ten times as many and the program takes about ten times as long to run. It is easy to see why we favor the program of Fig. 6-3 when the force is not velocity dependent. Through its use we can obtain greater accuracy with wider intervals and less computation time.

When we cannot find an exact solution to compare with our calculator solution, we need another technique to estimate the error. The simplest technique is to run the program for some appropriate interval width, then reduce the width significantly, rerun the program, and compare solutions. If they agree to a certain number of significant figures, then these figures constitute the accepted solution. If more accuracy is desired, the interval width is further reduced. Usually the width is divided by a number from 2 to 10. The next problem is an example of this procedure.

Problem 3. A 40 kg boy on a skateboard moves initially in the positive x direction and is subjected to the force

$$F = -5 - 0.5v$$

where v is the x component of his velocity. F is in newtons and v in meters per second. He starts with a speed of 5 m/s.

a. Show that
$$x = 1200(1 - e^{-0.0125t}) - 10t$$

and
$$v = 15e^{-0.0125t} - 10$$

satisfy v=dx/dt and F=mdv/dt, respectively, and that they produce the correct values for t=0. These are, in fact, the exact solutions.

b. Find his position and velocity at the end of 3 s, 6 s, 9 s, and 12 s by evaluating the exact analytic solutions given above.

c. Find the solutions using the program of Fig. 6-1. Do this 3 times, for interval widths of 1 s, 0.1 s, and 0.01 s, respectively.

By examining the results you can see that our prescription for assessing the error is rather conservative. After the $\Delta t=0.1$ s calculation, we would take x=47.1 m to be the position at the end of 12 s. These are the digits for which the $\Delta t=1$ s and the $\Delta t=0.1$ s calculations agree. Notice that acceptance of the next digit (x=47.14 m) brings the answer closer to the exact solution. At the completion of the Δt=0.01 s calculation, we would again take x=47.1 m to be the position at the end of 12 s. By our criteria, it would take another calculation, with a still smaller interval, to obtain another significant figure.

6.4   Exercises

1.   A person pushes a 150 kg crate along a rough floor. The force of the person on the crate is given by

$$F = 400e^{-0.15t}$$

where F is in newtons and t is in seconds. The exponential nature of the force takes into account the tiring of the person pushing. As long as the crate is moving there is also a frictional force of 175 newtons opposing the motion. How far is the crate pushed if it is initially at rest?

At small times the net force is in the same direction as the force of the person, in the positive x direction, say. The crate accelerates in that direction and its velocity is also in the positive x direction. At longer times, the force of the person weakens and the force of friction dominates. This slows the crate until it stops. Thereafter the force of friction has the same magnitude as the force of the person and the crate remains motionless. A good strategy to solve this problem is to make the erroneous assumption that the force of friction remains at 175 newtons. Now the velocity becomes zero at some time and then becomes negative. Find five values of the velocity in the neighborhood of this change in sign, fit to a fourth order polynomial and find the root. This is the time when the crate stops. Its position can be found by rerunning the program to integrate Newton's second law or by also fitting the values of the position to a polynomial and evaluating it for the root of the velocity polynomial. For purposes of integration, use Δt=0.05 s but display and use results only for every second.

2.   A 3.5 kg box is dragged along the ground by a constant force F=12 N acting horizontally. The box moves in the positive x direction. The ground is rougher

toward larger x and the coefficient of kinetic friction increases according to

$$\mu_k = 0.07\sqrt{x}.$$

a. Assume the box starts from rest at x=0 and use the program of Fig. 6-3 to find its position and velocity every second from t=0 to the time it comes to rest again. Calculate x and v for a few points for which v is negative, using the same equation for the force. This is not physically valid but it is mathematically sound and it allows you to interpolate to find the time the box comes to rest.

b. At what time does the box come to rest?

c. How far is the box dragged?

3. A gold nucleus repels a proton with a force given approximately by

$$F = 1.82 \times 10^{-26}/r^2$$

where F is in newtons and r is in meters. Here r is the distance between the nucleus and the proton. Suppose a gold nucleus is at the origin and a proton is shot directly at it along the x axis. When the proton is at $x=1.8 \times 10^{-8}$ m, it is moving toward the nucleus with a speed of 5×10^3 m/s. Assume the gold nucleus does not recoil but remains at the origin.

a. Generate a table which gives the position and velocity of the proton every 2×10^{-14} s from the time when it is at $x=1.8 \times 10^{-8}$ m to the time when it returns to approximately the same place. Use $\Delta t = 1 \times 10^{-15}$ s.

b. Find the time and position of closest approach by fitting the velocity and position data of part a to a fourth order polynomial.

c. Find the time and velocity when the proton is at $x=1.8 \times 10^{-8}$ m for the second time.

4. A 5 kg block starts from rest at the top and slides down an inclined plane which makes a 30° angle with the horizontal. The coefficient of kinetic friction increases with distance in such a way that

$$\mu_k = 0.9\sqrt{x}$$

where x is the distance along the plane, measured from the top.

a. Show that the position x of the block obeys

$$d^2x/dt^2 = g(\sin 30° - 0.9\sqrt{x}\cos 30°)$$

where x is measured from the top of the plane.

b. Use the program of Fig. 6-3 to find the position and velocity of the block every 0.1 s from the time block is released to the time the calculator produces a negative velocity. The negative velocity is not physically meaningful since the force law changes when the block stops. Obtain 3 significant figure accuracy.

c. Use graphical means to estimate the time the block stops and the distance it travels down the plane.

Chapter 7
NEWTON'S SECOND LAW IN TWO DIMENSIONS

The programs of the last chapter are modified for use when the particle moves in two dimensions. They are then applied to projectile motion with a drag force proportional to the square of the speed and to objects in circular and nearly circular orbit. These calculations supplement material in Chapters 4 and 6 of PHYSICS and Chapters 4 and 6 of FUNDAMENTALS OF PHYSICS as well as similar material in other texts. They provide additional demonstrations of the use of the second law in complicated physical situations. In particular they show very vividly that a two dimensional problem is simply two one dimensional problems solved simultaneously, even when the acceleration is produced by a centripetal force. The studies of projectile motion and satellites in circular orbit make excellent extended projects and a number of problems have been included with this in mind. If time is not available for a long project, one or two can be selected to illustrate the physics and numerical techniques.

7.1 Programs for Two Dimensional Motion

Newton's second law in two dimensions consists simply of the two one dimensional laws

$$F_x = m\, dv_x/dt \tag{7-1}$$

and

$$F_y = m\, dv_y/dt \tag{7-2}$$

where F_x and F_y are the x and y components, respectively, of the force and v_x and v_y are the x and y components, respectively, of the velocity.

The programs we present are much the same as the ones for one dimensional motion. The chief difference is that now we must keep track of two components of the position and two components of the velocity. In addition, each component of the force might depend on both components of the position and on both components of the velocity as well as on the time.

Fig. 7-1 gives the flow chart for a program which computes the force at the

75

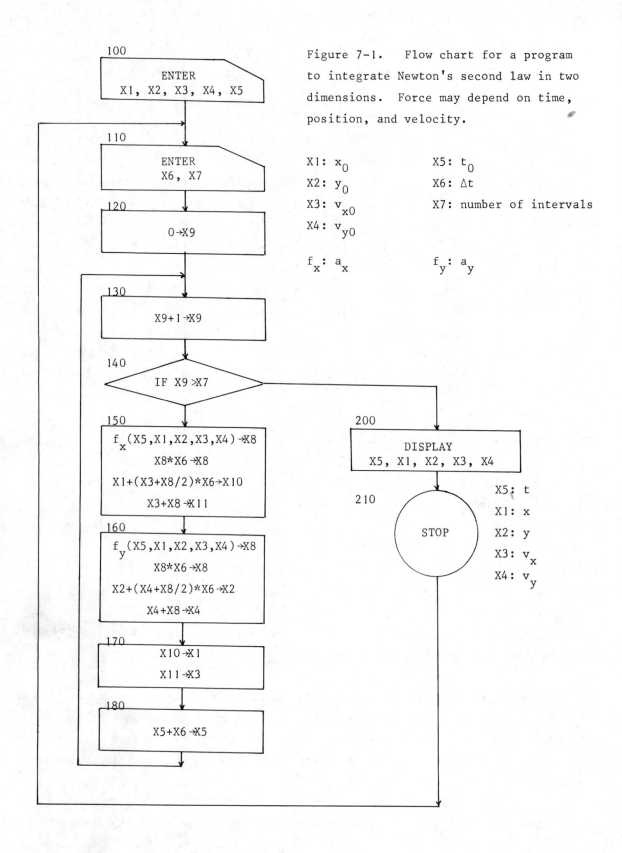

Figure 7-1. Flow chart for a program to integrate Newton's second law in two dimensions. Force may depend on time, position, and velocity.

X1: x_0 X5: t_0
X2: y_0 X6: Δt
X3: v_{x0} X7: number of intervals
X4: v_{y0}

f_x: a_x f_y: a_y

X5: t
X1: x
X2: y
X3: v_x
X4: v_y

beginning of each interval and uses the result, along with the position and velocity at the beginning of the interval, to calculate the position and velocity at the end of the interval. If the force is position or velocity dependent, the values of those quantities at the beginning of the interval are used to find the force for that time. This program is analogous to the program given in Fig. 6-1 for one dimensional motion.

Fig. 7-2 gives the flow chart for a program which is similar to the one in Fig. 6-3. The velocity components are calculated at the interval midpoints, using the force components at the interval boundary between. We expect this program to be more accurate than the first for the same interval width. On the other hand, this program cannot be used if the force is velocity dependent.

For both programs, the following storage allocations are made:

X1	x component of position,
X2	y component of position,
X3	x component of velocity,
X4	y component of velocity,
X5	time,
X6	interval width,
X7	number of intervals,
X8	x or y component of the force divided by the mass,

and X9   number of the current interval.

To save storage space, the same storage register, X8, is used for both components of the force divided by the mass. It is first used for the x component and, after that number is used to find the x component of the velocity and position, it is then used to store the y component.

In the flow chart, f_x is used to represent the x component of the force divided by the mass of the particle and f_y is used to represent the y component of the same quantity. As before, the user must supply the particular force function for the problem at hand. Both f_x and f_y may depend on both components of the position and on both components of the velocity when the program of Fig. 7-1 is used. For the program of Fig. 7-2, f_x and f_y may depend on both components of the position. In either case, they may also depend on time. Values of x and v_x are stored in X10 and X11 respectively until y and v_y are computed. In all other respects, the programs are like the analogous

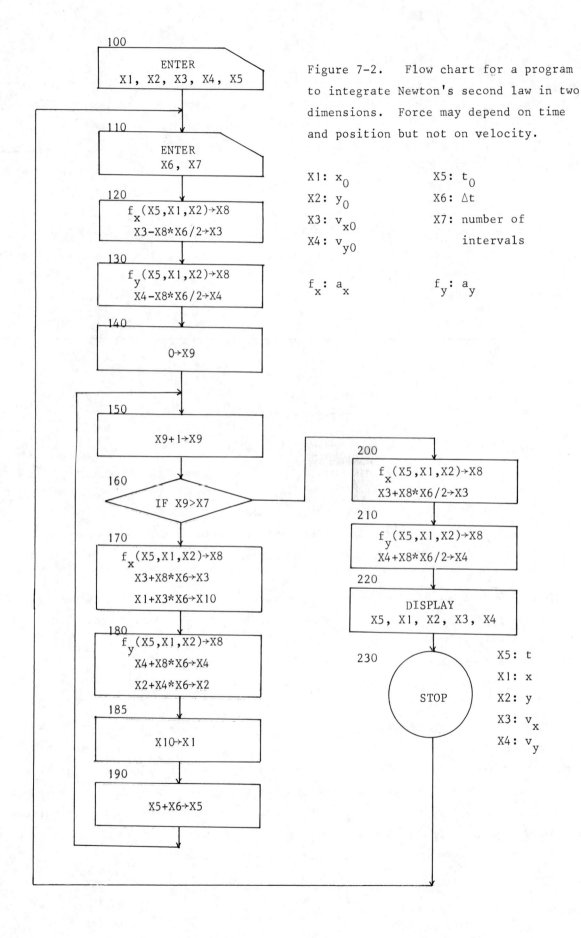

Figure 7-2. Flow chart for a program to integrate Newton's second law in two dimensions. Force may depend on time and position but not on velocity.

X1: x_0 X5: t_0
X2: y_0 X6: Δt
X3: v_{x0} X7: number of
X4: v_{y0} intervals

f_x: a_x f_y: a_y

X5: t
X1: x
X2: y
X3: v_x
X4: v_y

programs of the last chapter.

If the same interval width and number of intervals are used throughout the running of the program, both programs should be modified so that the machine returns to line 120 after the STOP instruction.

7.2 Projectile Motion with a Drag Force

For most projectiles and falling bodies, the drag force is proportional to the square of the speed. General solutuions to Newton's second law equation do not exist for forces of this type and it is necessary to resort to numerical techniques.

The drag force can be written

$$\vec{F}_D = -bv\vec{v} \qquad (7\text{-}3)$$

where b is a constant of proportionality, $\vec{v}$ is the velocity of the projectile, and v is the magnitude of $\vec{v}$. Note that $-v\vec{v}$ is a vector with magnitude v^2 and is directed opposite to the velocity.

For a projectile moving in the x,y plane and subject to a gravitational force of magnitude mg in the negative y direction, the equations of motion are

$$m\, dv_x/dt = -bv_x(v_x^2 + v_y^2)^{\frac{1}{2}} \qquad (7\text{-}4)$$

and

$$m\, dv_y/dt = -mg - bv_y(v_x^2 + v_y^2)^{\frac{1}{2}}. \qquad (7\text{-}5)$$

There is a terminal velocity. After a long time has elapsed, the acceleration vanishes and the velocity becomes

$$\vec{v}_{term} = -\sqrt{mg/b}\,\hat{j} \qquad (7\text{-}6)$$

The x component of the velocity vanishes and the y component becomes constant. Of course, the projectile may hit the ground or some other object before it reaches terminal velocity.

For large values of b, we expect v_x to vanish rapidly and the projectile to become a falling body. Instead of a parabola, symmetric about the highest point, the trajectory is blunted in the forward direction and the projectile makes little progress in that direction as it falls.

The following problems are designed to help you make a study of this type motion. They do not answer all the questions that come to mind but, after working on them, you should be able to design programs to study any aspect of the motion. Some of those given here take several hours to run on a hand held calculator.

In order to show the influence of the drag force on projectile motion, we present the following problem, in which projectile motions without and with air resistance are compared.

Problem 1. A projectile is launched from the origin with an initial speed of 150 m/s at an angle $40°$ above the horizontal.
a. Assume zero air resistance and use the program for velocity independent forces (Fig. 7-1) to plot the trajectory, y vs. x. Take the interval Δt to be 0.2 s and plot the position of the projectile every 2 s from the time of launch to the time its y coordinate becomes negative. Save the values of the velocity components for use in parts c and d.
b. Assume $b/m = 2 \times 10^{-3}$ m^{-1} and repeat part a using the program of Fig. 7-2 with $\Delta t = 0.05$ s. Use the same graph paper. Save the values of the velocity components for parts c and d.
c. Plot, on another graph, two graphs showing v_x vs. t for motion without and with air resistance, respectively.
d. Plot, on another graph paper, two graphs showing v_y vs. t for motion without and with air resistance, respectively. Use Eqn. 7-6 to calculate the terminal velocity and mark its value on the graph.

The time interval suggested for part b is rather large for this calculation and there is error in the third significant figure for some of the results. If your machine is fast you might want to reduce Δt by a factor of 2 or more.

The graphs drawn in connection with problem 1 show many of the typical features of projectile motion. It is immediately obvious that air resistance shrinks the trajectory.

For the value of b used, the highest point is much lower, the range is much less, and the time of flight is much less than for the zero resistance trajectory.

Notice also that the trajectory for b=0 is symmetric about the highest point. Points on the graph to the right of the highest point have the same y coordinate as analagous points the same distance to the left. On the other hand, the trajectory for $b/m=2\times10^{-3}$ m^{-1} shows the typical blunt-nosed shape for a projectile subjected to a drag force.

The velocity curves show the x component approaching zero and the y component approaching the terminal velocity as time goes on.

e. Assume the two projectiles are launched simultaneously and use the graphs to determine if there are any times for which the y component of the velocity for the projectile subjected to air resistance is greater than the y component of the velocity for the projectile which is not.

f. Suppose the impact point is not y=0. Is it possible to fire the projectiles onto a plateau or into a valley (at the same firing speed and angle) so that the y component of velocity, on impact, has greater magnitude for the projectile subjected to air resistance than for the one which is not?

The following problems are included so that you can make a more intensive study of projectile motion with a drag force. Most of these problems take considerable time on all but very fast machines. Nevertheless, they should give you a good understanding of this type motion and are well worth the effort.

<u>Problem 2</u>. A 0.05 kg pellet has a drag coefficient $b=6.3\times10^{-4}$ kg/m and is launched from the origin at $30°$ to the horizontal. For each of the following initial speeds, plot the trajectory y vs. x. Take the integration interval to be $\Delta t=0.01$ s and plot points every 0.5 s from t=0 to the time when y becomes negative.
a. 20 m/s.
b. 50 m/s.
c. 100 m/s.

Notice the lack of symmetry about the highest point. As the initial speed is made greater, the value of x for the highest point is larger and the range is also greater. However, the range does not increase as much as the x coordinate of the highest point and the curve has a more blunt looking appearance for higher initial speed.

Problem 3. The pellet of problem 2 is launched from the origin with a speed of 50 m/s. For each of the following firing angles, plot the trajectory, y vs. x. Take the integration interval to be Δt=0.01 s and plot points every 0.5 s from t=0 to the time when y becomes negative. For later use, calculate 2 points for which y is negative.
a. $30°$ above the horizontal (see problem 2).
b. $35°$ above the horizontal.
c. $40°$ above the horizontal.
d. $45°$ above the horizontal.
e. $50°$ above the horizontal.

Notice how the shape of the trajectory changes with firing angle and also notice that the longest range occurs for a firing angle less than $45°$.

Problem 4. For each of the firing angles of problem 3, calculate the range. Do this by fitting the points near y=0 to a polynomial and finding the root. Then fit the range as a function of firing angle to a polynomial and find the firing angle for maximum range by solving for the root of its derivative. How does the maximum range compare with the maximum range for the same initial speed but zero air resistance? The firing angle is then $45°$.

Problem 5. To see how the drag coefficient influences the trajectory, consider the case when the pellet of problem 2 is launched with an initial speed of 50 m/s at a firing angle of $30°$. Take $b=6.3\times10^{-3}$ kg/m, Δt=0.01 s, and find the position every 0.5 s from t=0 to the time when the y coordinate becomes negative. Compare the trajectory with that of the projectile of problem 2, part c. Plot both trajectories on the same graph.

7.3 Circular and Nearly Circular Orbits

We consider situations in which a mass m follows a circular orbit centered at the origin. It is pulled toward the origin by a force $\vec{F}$ whose magnitude may depend on the distance from the origin to the mass.

We start the mass with some initial velocity $\vec{v}$, at some distance R from the origin. The velocity is tangent to the circle of radius R centered at the origin. If the force is radially inward and its magnitude is given by

$$F = mv^2/R \tag{7-7}$$

then the orbit is circular and the speed remains constant.

In the first problem, we consider an object executing uniform (constant speed) circular motion under the influence of a force with constant magnitude. The value of doing this problem arises from the fact that we use exactly the same program as we use for projectile motion or any other Newton's second law problem in two dimensions.

Problem 1. A 0.2 kg block is tied to a rope of length 1.6 m and caused to move at 3 m/s in a circle with radius equal to the length of the rope.
a. What is the force of the rope on the block? (ans: 1.125 N)
b. Show that, in general

$$F_x = -1.125 \, x/(x^2 + y^2)^{\frac{1}{2}}$$

and

$$F_y = -1.125 \, y/(x^2 + y^2)^{\frac{1}{2}}$$

    are the components of the force when the coordinates of the block are x and y. Here the force components are in newtons and the coordinates are in meters.

c. Start the block at x=1.6m, y=0 with velocity components $v_{0x}=0$, $v_{0y}=3$ m/s. Use the program for velocity independent forces (Fig. 7-2), with a time interval $\Delta t=0.05$ s, and calculate the position and velocity of the block for every 0.5 s from t=0 until it completes one revolution.

d. For each of these times, calculate $R=(x^2 + y^2)^{\frac{1}{2}}$ and $v=(v_x^2 + v_y^2)^{\frac{1}{2}}$ to see if they are constant (within the accuracy of the program).

e. Plot the trajectory, y vs. x.

The magnitude of the gravitational force of the earth acting on a satellite of mass m, a distance R from the earth's center, is

$$F = G \frac{Mm}{R^2}$$

where G is the universal gravitational constant (6.67×10^{-11} m^3/s$^2 \cdot$kg) and M is the mass of the earth (5.98×10^{24} kg).

In the next problem, we consider a satellite in orbit around the earth. We set up conditions for a circular orbit but then take into account a drag force such as the atmosphere might exert on the satellite. We use the calculator to "watch" it spiral in toward the earth. In truth, the drag coefficient we use is too large to be realistic. The effect has been exaggerated so that the program does not take an unreasonable amount of time to run.

Problem 2. Consider a satellite which, if there were no resistive force, would move in a circular orbit with a radius which is twice the radius of the earth (radius of the earth = 6.378×10^6 m).

a. What is the speed of the satellite?

b. Show that the x and y components of the gravitational force acting on the satellite are

$$F_x = -GMmx/R^3$$

and

$$F_y = -GMmy/R^3,$$

respectively. R is the orbit radius, x and y are the satellite's coordinates.

c. Set up appropriate initial conditions (position and velocity) for a circular orbit. For example, put the satellite on the x axis and give it an initial velocity in the y direction. Use the program of Fig. 7-2, with $\Delta t = 100$ s, to calculate the position and velocity for every 1000 s. Also calculate $R = (x^2 + y^2)^{\frac{1}{2}}$ and $v = (v_x^2 + v_y^2)^{\frac{1}{2}}$ to see if they are constant. They should flucuate only in the fourth significant figure. Make a plot of the orbit for one revolution.

d. Now include a drag force proportional to the square of the speed. Show that

$$F_x = -GMmx/R^3 - bv_x(v_x^2 + v_y^2)^{1/2}$$

and

$$F_y = -GMmy/R^3 - bv_y(v_x^2 + v_y^2)^{1/2}$$

are the x and y components, respectively, of the force.

e. Use the program for velocity dependent forces (Fig. 7-1) to make a plot of the orbit, y vs. x. Use $\Delta t=10$ s, take $b/m=1.2\times 10^{-8}$ kg/m, and plot points for every 500 s. Stop 2 points beyond the crash, which occurs when the satellite is one earth radius from the center of the earth. There is error in the third figure.

f. How does the speed change? For each of the points of part e, calculate $v=(v_x^2 + v_y^2)^{1/2}$ and plot v as a function of time.

Can you explain the change in speed in terms of the forces acting? A force which is perpendicular to the velocity changes the direction of the velocity but does not change its magnitude. To increase the speed, there must be a non-vanishing component of the total force in the same direction as the velocity; to decrease the speed, there must be a non-vanishing component of the total force in the direction opposite to the velocity.

g. When does the satellite crash into the surface of the earth? Calculate $R=(x^2 + y^2)^{1/2}$ for each of the points of part e, use appropriate values to find the coefficients for a polynomial approximation to R(t), and use a root finding program to find the time for which R is equal to the earth's radius. Also find the speed of the satellite at the time of the crash.

h. If the drag force were suddenly reduced to zero near the end of the first quarter revolution, would the satellite enter a circular orbit?

Save the answers to these questions. They will be used later for a problem in Chapter 8.

Chapter 8

INTEGRATION, WITH APPLICATIONS TO THE CALCULATION OF WORK

The first section of this chapter contains a program to evaluate the definite integral of a function. In subsequent sections, it is used to calculate the work done by a force. A numerical demonstration which shows the physical meaning of a potential energy function is included and the meaning is stressed by a calculation of work for both a conservative and a non-conservative force. In later chapters the program is used to evaluate other integrals. This chapter supplements the material in Chapters 7 and 8 of PHYSICS, chapters 7 and 8 of FUNDAMENTALS OF PHYSICS, and similar material in other texts. Its purpose is to sharpen the concepts of work and potential energy through detailed calculation.

8.1 Integration by Simpson's Rule

In this section we consider the numerical evaluation of definite integrals of the form

$$I = \int_{x_0}^{x_f} f(x) \, dx \qquad (8-1)$$

where the function $f(x)$ is given. The technique is to divide the range into many segments, fit a different polynomial in each segment, integrate the polynomials, and sum the contributions of all the segments.

It should be clear that, in general, the higher the order of polynomial used, the fewer the number of segments one needs. On the other hand, a high order polynomial requires a large number of registers to store the coefficients and may result in loss of significance. One must make a compromise. For our purposes, a second order polynomial, fit to 3 points, seems to be a good choice.

The computational scheme we describe is known as Simpson's rule for evaluating integrals. To apply the scheme, the range from x_0 to x_f is divided into N segments, each of width h, and the points which mark the boundaries of the segments are labelled x_0, x_1, x_2, ... x_N where $x_1=x_0+h$, $x_2=x_0+2h$, ... $x_N=x_0+Nh=x_f$.

We fit a second order polynomial to the 3 points x_0, x_1, x_2; another to the points x_2, x_3, x_4; another to the points x_4, x_5, x_6; and so on until we finish with the last triplet of points, x_{N-2}, x_{N-1}, x_N. It is clear that N must be even for this scheme to work. The central point of each triplet is a point labelled with an odd number and x_N must be the third point in a triplet.

For ease in writing what follows, let f_i be the value of the function for $x=x_i$. That is $f_0=f(x_0)$, $f_1=f(x_1)$, ... $f_N=f(x_N)$.

First consider the interval from x_0 to x_2. The second order polynomial which correctly reproduces the value of the function at the three points x_0, x_1, and x_2 is

with
$$P = A_0 + A_1(x-x_0) + A_2(x-x_0)^2$$
$$A_0 = f_0,$$
$$A_1 = -(f_2 - 4f_1 + 3f_0)/2h,$$
and
$$A_2 = (f_2 - 2f_1 + f_0)/2h^2. \qquad (8-2)$$

The contribution of this interval to the integral is

$$\int_{x_0}^{x_2} P\, dx = \int_{x_0}^{x_2} \left[A_0 + A_1(x-x_0) + A_2(x-x_0)^2\right] dx$$
$$= A_0(x_2-x_0) + \tfrac{1}{2} A_1(x_2-x_0)^2 + \tfrac{1}{3} A_2(x_2-x_0)^3. \qquad (8-3)$$

Now substitute the expressions for the coefficients in the appropriate places and also make the substitution $x_2-x_0=2h$. Then the contribution becomes

$$\int_{x_0}^{x_2} P\, dx = \tfrac{h}{3}(f_0 + 4f_1 + f_2). \qquad (8-4)$$

Similar results hold for the other intervals. Each contribution is, in general, h/3 multiplied by the sum of the function at the first point, 4 times the function at the midpoint, and the function at the third point.

Now add the contributions. Notice that every point with an even numbered label, except 0 and N, enters twice: once as the end point of an interval and once as the initial point of the next interval. f_0 and f_N each enter once. Every point with an odd numbered label is multiplied by 4. We write for the sum of the contributions

$$I = (h/3) \left[(f_0 - f_N) + 2(f_2 + f_4 + f_6 + \ldots + f_N) \right.$$
$$\left. + 4(f_1 + f_3 + f_5 + \ldots + f_{N-1}) \right]. \quad (8-5)$$

The second set of parentheses contains the contributions of all points with even labels, including the point N but excluding the point 0. This sum is multiplied by 2. Notice that this procedure counts f_N twice as much as its contribution to the integral should be. The third set of parentheses contains the contributions of all the points with odd labels and the sum is multiplied by 4. The first set of parentheses contains the contribution of f_0. It is here that the contribution of f_N is subtracted to make up for being included in the second set of parentheses.

The terms are grouped in this manner for ease in programming. Points are paired and, each time around the program loop, two contributions are found. We choose to find the contributions of f_1 and f_2 during the first traversal, the contributions of f_3 and f_4 during the second traversal, and so on. During the final traversal, we find the contributions of f_{N-1} and f_N. This means we traverse the loop N/2 times.

Fig. 8-1 shows the flow chart for a program to evaluate Eqn. 8-5. At line 100, X1 is the initial point (x_0), X2 is the final point ($x_f = x_N$), and X3 is the number of segments N. At line 110, the width of the interval is calculated and, at 120, half the number of segments is computed and stored in X3. This number is needed since the loop is traversed N/2 times.

In what follows, X1 contains the value of x used to evaluate the function, X6 contains the sum of the values of the function at the odd labelled points ($f_1 + f_3 + \ldots + f_{N-1}$), X7 contains the sum of the values of the function at the even labelled points ($f_2 + f_4 + \ldots + f_N$), and X8 is a counter which counts the number of times the loop is started. At line 140, X6, X7, and X8 are set equal to 0 prior to entering the loop.

At line 150, X8 is incremented by 1 and, at line 160, its value is compared with that of X3 to see if the loop should be traversed again. If it should, the machine proceeds to line 170, where X1 (x) is incremented by X4 (h). As we shall see, this always evaluates x for one of the points with an odd numbered label. The first time the loop is entered, X1 contains x_0 and it is incremented to x_1. On subsequent

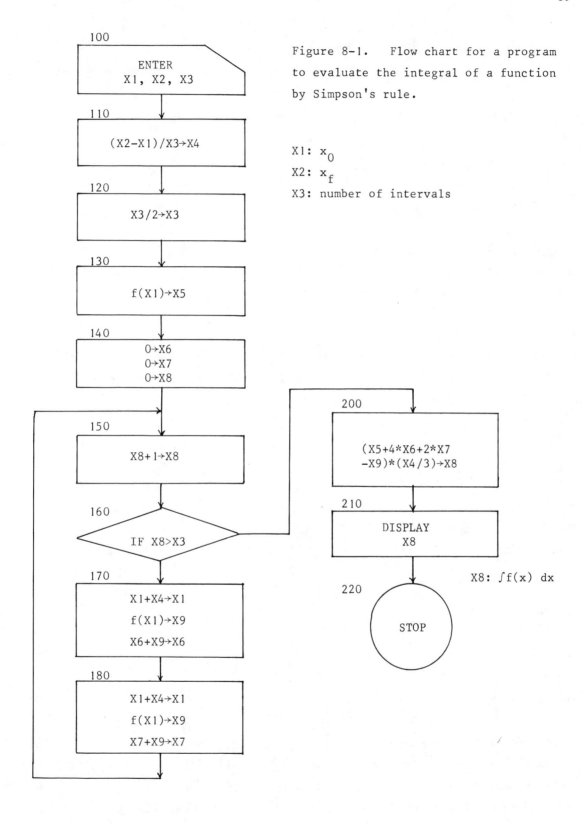

Figure 8-1. Flow chart for a program to evaluate the integral of a function by Simpson's rule.

X1: x_0
X2: x_f
X3: number of intervals

X8: $\int f(x)\, dx$

traversals, it takes on the values x_3, x_5, ...

The function is then evaluated for X1, its value is added to the number in X6, and the sum is stored back in X6. Thus X6 collects the sum $f_1 + f_3 + \ldots$.

At line 180, X1 is again incremented by X4 (h). Since x was a point with an odd numbered label, it is now a point with an even numbered label. The function is evaluated for X1, its value is added to the number in X7, and the sum is stored back in X7. Thus X7 collects the sum $f_2 + f_4 + \ldots$.

The next time around the loop, the machine reaches line 150 with X1 containing the coordinate of a point with an even numbered label so, after it is incremented at line 170, it contains the coordinate of a point with an odd numbered label, as it should.

After the loop is traversed N/2 times, the program goes to line 200, where the various pieces are put together. Note that X9 still contains f_N. Finally, the value of the integral is displayed at line 210.

8.2 Calculation of Work for Motion in One Dimension

In this section, we consider a particle which moves along the x axis and calculate the work done by a force with x component denoted by F. There may be other forces acting on the particle and it is the total force which determines the acceleration of the particle. Nevertheless, we wish to calculate only the work done by the force we select. Specifically, the work done by the force as the particle moves from x_0 to x_f is given by

$$W = \int_{x_0}^{x_f} F \, dx \tag{8-6}$$

and it is this integral we wish to evaluate.

There are two classes of forces which need to be distinguished: forces which depend only on the position of the particle are placed in the first group, while forces which depend on the time or on the velocity of the particle are placed in the second group. For placement in the second group, it is immaterial whether or not the force also depends on the position of the particle. We employ different computational schemes to calculate the work done by these two types of forces.

There is an exception to the above classification. There are some forces whose magnitudes depend only on the position of the particle but whose directions depend on the velocity. One example is a force of friction with constant magnitude. It is always in a direction opposite to the direction of the velocity and when the velocity changes direction so does the force. Such forces are, strictly speaking, members of the second group, but it is usually convenient to modify the computational scheme designed for forces of the first group so that the work done by these forces can be calculated using the same scheme. We deal with these forces separately.

If the force depends only on position, the technique and program of Fig. 8-1 can be used without modification. The object is to calculate the work done by the force as the particle moves from x_0 to x_f. We enter x_0 into X1, X_f into X2, and the number of segments into X3. We know that both accuracy and computation time increase with the number of segments.

Notice that this computation scheme gives the correct sign for the work. If F is positive over the entire range and x_f is more positive than x_0, then the computed work is positive as it should be. On the other hand, if F is positive but x_0 is more positive than x_f, then the computed work is negative, again as it should be.

There is another important point to be noted. Suppose the particle goes from x_0 to a point x_A, beyond x_f, then reverses direction and goes back to x_f (see Fig. 8-2).

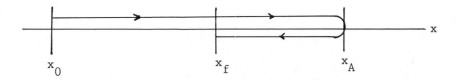

Figure 8-2. An example of a path of integration used to calculate the work done by a force.

During the course of its motion, the particle gets to some parts of the x axis more than once. To calculate the work, we should, in principle, follow the path of the particle and integrate from x_0 to x_A and then from x_A to x_f. This requires running the program twice since there are different limits on the integration for the two segments. However, this process is not required for the situation being discussed: the motion is in one dimension and the force depends only on the position of the particle. Under these conditions, the work done by the force as the particle moves from x_f to x_A (going to the right in Fig. 8-2) exactly cancels the work done by the same force as the particle goes from x_A to x_f (going to the left). On the two occasions when the particle is at any point, the force is the same (it depends only on the position) but the limits of the integration are reversed.

As a result of this discussion, we conclude that we need tell the program only x_0 and x_f to calculate the work. How many times the particle reverses direction does not matter as long as it finally gets to x_f.

We now discuss the case for which the magnitude of the force depends only on the particle's position but the direction of the force depends on the direction of the velocity.

In this instance we must be careful if the particle reverses its direction of travel and traverses some parts of the path more than once. Now the sign of the force component is reversed along with the limits of integration so the contribution of the segment has the same sign for each traversal and no cancellation occurs. To evaluate the work, it is now important to follow the particle from x_0 to x_A and then back to x_f. The program must be run twice and the results added.

For forces of friction, the path must be split into a number of segments so that the direction of travel is the same throughout each segment. The boundaries of the segments are the places where the particle reverses direction. The program is then run once for each segment.

8.3 Time and Velocity Dependent Forces

If the force depends explicitly on the time or on the velocity of the particle, the computing and data handling problem is somewhat more complicated. In order to know the force acting on the particle when it is at the position x, one must know the details of the motion: the time when it is at x and the velocity when it is at x. For many cases, this means the integration of Newton's second law. Incidently, the force used in Newton's second law is the total force on the particle and not necessarily the force whose work we are computing.

The result of a numerical integration of the second law is usually a table containing values of x and v for various times. In what follows, we assume these values are given at uniform intervals of time. Because the input data for the calculation of work is usually in this form, it is convenient to modify the technique used so that we use time, rather than position, as the variable of integration.

To do this, we recognize that dx=v dt where v is the x component of the particle's velocity at time t. The work done by the force F from time t_0 to time t_f can then be written

$$W = \int_{t_0}^{t_f} F(x,v,t) \; v \; dt. \tag{8-7}$$

To evaluate this integral we again use Eqn. 8-5 but now h is a time interval instead of a displacement interval and it is the function Fv which is evaluated at the various points rather than just F alone. The same program (Fig. 8-1) can be used. Now t_0 is entered into X1 and t_f is entered into X2. At lines 130, 170, and 180 one must remember to evaluate Fv as the function f.

This technique automatically follows the particle as it moves along its path. If the particle traverses some part of the x axis more than once, all the contributions are automatically included in W.

Typically, the input to this program is obtained as output from one of the programs used to integrate Newton's second law. This source of data means that care must be exercised in choosing the parameters for the integration of the second law. One needs an even number of segments (an odd number of points, including the initial point), close enough together to give acceptable accuracy when the work is calculated. For hand held calculators, this poses somewhat of a problem. Since there are not enough storage registers to store x and v for all the required values of time, these numbers must be written down and then, when the work program is run, must be entered at appropriate times during the run. Lines 130, 170, and 180 of the work program must be modified to receive data. In particular, each of these lines must contain an ENTER statement to read the appropriate values of x and v into storage. These values are then used to calculate Fv.

8.4 Work for Motion in Two Dimensions

When the position of the particle changes by the infinitesimal vector $d\vec{r}$, the force $\vec{F}$ does work dW, given by

$$dW = \vec{F} \cdot d\vec{r} \qquad (8-8)$$

The displacement $d\vec{r}$ has magnitude equal to the distance travelled along the path and is a vector tangent to the path of the particle. For a path of finite length, we add the contributions of a large number of segments and write, in the limit of an infinite number of segments

$$W = \int_a^b \vec{F} \cdot d\vec{r} \ . \qquad (8-9)$$

This gives the work done by the force $\vec{F}$ as the particle moves along its trajectory from point a to point b.

We again deal with two distinct cases. For the first, the position and velocity of the particle are known functions of time. They may be given in analytic or tabular form. For the second case, the position and velocity are not known but the force is a function only of the position of the particle and the path of the particle is known.

If the position and velocity of the particle are given as functions of time, it is usually convenient to rewrite the integrand so that the work is calculated using time as the variable of integration. We observe that $d\vec{r} = \vec{v} \, dt$, where $\vec{v}$ is the velocity of the particle, and write

$$W = \int_{t_0}^{t_f} \vec{F} \cdot \vec{v} \, dt \ . \qquad (8-10)$$

This is the work done by the force $\vec{F}$ during the time from t_0 to t_f as the particle moves along its trajectory. This expression is really quite similar to Eqn. 8-7 for the one dimensional case. The difference is that we now use $\vec{F} \cdot \vec{v} = F_x v_x + F_y v_y$ as the integrand instead of $F_x v_x$ (which we wrote Fv in the last section).

The program of Fig. 8-1 can easily be adopted to calculate the work done by a force when the particle moves in two dimensions. The function f which appears at lines 130, 170, and 180 is $\vec{F} \cdot \vec{v}$, X1 is the initial time t_0, X2 is the final time t_f, and X3 is still the number of intervals N.

One must supply information to calculate $\vec{F} \cdot \vec{v}$ for the times t_0, t_0+h, t_0+2h, ... t_0+Nh. Usually the position and velocity of the particle are known for selected times as a result of a previous integration of Newton's second law. In this case, they can be supplied by entering the appropriate values for each time as needed by the work calculation program. Part of the "calculation" at lines 130, 170, and 180 consists of ENTER statements.

In all two dimensional cases discussed so far, the program of Fig. 8-1 is used with $f=\vec{F} \cdot \vec{v}$ and the intervals used are time intervals. Even if the force is independent of the velocity, this procedure is used if the velocity can be found as a function of time by some means. $d\vec{r}=\vec{v}\,dt$ is usually the most convenient way to find $d\vec{r}$.

There are, however, some circumstances in which $\vec{F}$ depends only on the position of the particle, the path is known, but the position of the particle along the path is not known as a function of time. The velocity is not known and the integral

$$W = \int_a^b \vec{F} \cdot d\vec{r}$$

must be dealt with directly and not converted to an integral over time.

The problem now centers around the specification of $d\vec{r}$. We must find a vector which is tangent to the path and has magnitude equal to the path length of a segment. To do this we need to examine how the path itself is described.

The path may be specified by giving the y coordinate of a point on the path as a function of the x coordinate of that point. We assume in what follows that

$$y = g(x) \qquad (8-11)$$

is a known function which describes the relationship between the coordinates of points on the path. A plot of this function is a drawing of the path. If y is known only at a set of discrete values of x, then the function g(x) might be a polynomial fit to those points. In any event, it is important for what follows that we be able to evaluate the derivative dy/dx.

We are now ready to find $\vec{dr}$. First we note that $\vec{dr} = dx\hat{i} + dy\hat{j}$ and that dx and dy are related to each other. If $\vec{dr}$ is tangent to the path, then dy/dx is the slope of the path and

$$\vec{dr} = \left[\hat{i} + (dy/dx)\hat{j}\right] dx. \qquad (8-12)$$

The derivative here is the derivative of Eqn. 8-11.

For most calculations of work, it is convenient to use x as the variable of integration. The integral then takes the form

$$W = \int_{x_0}^{x_f} \left[F_x + (dy/dx)F_y\right] dx. \qquad (8-13)$$

The program of Fig. 8-1 can again be used if the following interpretations are made. X1 is the x coordinate of the initial point (x_0), X2 is the x coordinate of the final point (x_f), and X3 is the number of intervals to be used in the integration. The function f which is evaluated at lines 130, 170, and 180 is

$$f = F_x + (dy/dx)F_y. \qquad (8-14)$$

Each of the components F_x and F_y of the force may be functions of both x and y. The x coordinate is chosen by incrementing the previous value of x by the interval width h. The y coordinate is found by the evaluation of y=g(x), and the slope (dy/dx) of the path is found by the evaluation of dg/dx.

We assume the path is specified by the functional relationship y=g(x) with the derivative dy/dx=g´(x) and give the program modifications necessary to perform the calculation of the work done by $\vec{F}$. The series of steps at line 130 becomes

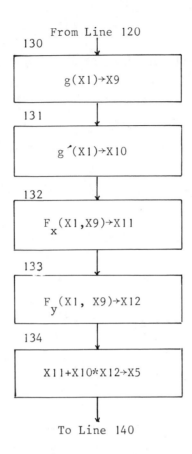

Here F_x and F_y represent the x and y components, respectively, of the force. If the path, the force, or both are given as polynomials, it is worthwhile to enter the coefficients at the beginning of the program, after line 100, for example. Then at 130, 131, 132, and 133 the coefficients are used to evaluate the functions.

Replacement statements are also needed for lines 170 and 180. In each case, they are identical to the ones given above, except that the final result is placed in X9. The first statement in each box (X1+X4→X1) is retained in both cases as is the last statement (X6+X9→X6 or X7+X9→X7).

There is one situation for which this program does not work. When the path is parallel or nearly parallel to the y axis, then dy/dx is extremely large and h must be taken to be extremely small. For paths which are nearly parallel to the y axis, it is usually convenient to integrate in y rather than in x. In this case we write

$$\vec{F} \cdot d\vec{r} = \left[(dx/dy)F_x + F_y\right] dy. \tag{8-15}$$

The same program is used. Now X1 is the starting value of y, X2 is the end value of y, and X3 is the number of intervals to be used. The path is specified by giving $x=g(y)$ and the derivative $dx/dy=g'(y)$.

If the same path is nearly parallel to both the x and y axes (at different places), then the calculation of work must be done in more than one part and the contributions added.

8.5 Exercises

1. A 2 kg block starts from rest at $x=0$ and moves along the x axis. The total force acting on it is given by $\vec{F} = (8 + 12t^2)\hat{i}$ where $\vec{F}$ is in newtons and t is in seconds.
   a. Use analytical means to show that
   $$v(t) = 4t + 2t^3$$
   and
   $$x(t) = 2t^2 + 0.5t^4.$$
   b. The force $\vec{F}_1$ (one of the forces which go to make up $\vec{F}$) is given by $\vec{F}_1 = 3t^2\hat{i}$. Find the work done by $\vec{F}_1$ during the first 3 seconds of the motion. Do this by direct analytic integration of $\vec{F}_1 \cdot \vec{v}\, dt$ and by use of the program of Fig. 8-1. Use an integration interval so that you obtain agreement to 3 significant figures.

2. A 2 kg block moves along the x axis, subjected to the force $\vec{F} = (6 - 4x)\hat{i}$. $\vec{F}$ is in newtons and x in meters.
   a. How much work is done by this force as the block moves from $x=0$ to $x=1$ m? Answer by direct analytical integration and by numerical integration, using the program of Fig. 8-1 with N=32.
   b. How much work is done by the force as the particle moves from $x=0$ to $x=5$ m? Answer by analytical integration and by numerical integration.
   c. If $\vec{F}$ is the total force acting on the block, then the work that it does is equal to the change in the kinetic energy of the block. Suppose the block starts with a speed of 6 m/s. What then is its speed when it is at $x=1$ m? at $x=5$ m?

3. The position as a function of time for a 25 kg crate sliding across the floor (along the x axis) is $x(t) = 3e^{-0.6t}$, in meters with t in seconds.
    a. Numerically evaluate the integral for the work done by the total force during the first 2 seconds. You must find $v(t)$ and $F(t)$ by differentiating $x(t)$ and using F=ma. Select an integration interval so that you expect three significant figure accuracy in your answer.
    b. Use $K = \frac{1}{2}mv^2$ to evaluate the kinetic energy at t=0 and at t= 2 s. Compare the change with your answer to part a. Did you achieve 3 significant figure accuracy?

In section 3.4, we studied one dimensional motion in the earth's gravitational field (force mg directed downward) with a force of air resistance proportional to the velocity. For this motion

$$y(t) = \frac{1}{b}\left(\frac{g}{b} + v_0\right)(1 - e^{-bt}) - \frac{gt}{b}$$

and

$$v(t) = \left(\frac{g}{b} + v_0\right)e^{-bt} - \frac{g}{b},$$

where the y axis is taken to be positive in the up direction, v_0 is the initial velocity, b is the drag coefficient, and g=9.8 m/s^2.

One dimensional motion with a drag force provides a non-trival example for the calculation of work done by a velocity dependent force. It also provides an interesting example of the work energy theorem.

4. A 0.5 kg ball is thrown nearly straight upward from the edge of a cliff with an initial speed of 25 m/s. On the way down, it misses the cliff edge and continues to fall to the canyon floor 350 m below. Take b=0.1 s^{-1} and use your answers to problem 5 of section 3.4.
    a. Use 50 intervals (in time) from the time the ball is thrown to the time it hits the canyon floor and calculate the work done by the force of air resistance, F=-mbv.
    b. Calculate the work done by gravity during the same time interval. This is, of course, just mgh where h is the distance the final point is below the initial point. h=+350 m for this problem.

c. Calculate the change in the kinetic energy. To how many significant figures do your answers support the work energy theorem?

5. Consider the falling satellite of problem 2 in section 7.3. Take the mass of the satellite to be 1 kg and use the data you generated when you worked that problem to answer the following questions.
    a. Fit v^3 to a fourth order polynomial using points at t=0, 500 s, 1000 s, 1500 s, and 2000 s. Use this polynomial to calculate the work done by the drag force during the first 2000 s. Also fit $(\vec{R}\cdot\vec{v})/R^3$ to a polynomial, using the same points, and calculate the work done by the force of gravity during the same interval.
    b. Repeat the calculation for the time interval from 2000 s to 4000 s. Use the points at t=2000 s, 2500 s, 3000 s, 3500 s, and 4000 s to fit the polynomials.
    c. Use the numerical results to check the work energy theorem for the first 2000 s and for the second 2000 s. We expect error in the third figure.
    d. In section 7.3, we asked why the speed of the satellite starts to increase after the first 2000 s or so. Do your results for the above calculations substantiate your previous answer? Can you explain why the speed decreases at first, then increases? What force is causing the change in each case?

6. In exercise 3 of section 6.4, you obtained data for a "bullet" proton impinging on a gold nucleus at the origin. Use the data to calculate the work done on the proton from t=0 to $t=16\times10^{-14}$ s. Do this by numerically evaluating the integral of Fv. Compare the answer with the change in the kinetic energy.

8.6 Potential Energy Functions

If a force is conservative, then the work it does as the particle moves between any two points is the same regardless of the path taken by the particle. This property of the force allows us to calculate the work in a special way, not applicable to non-conservative forces. We can assign a number to each point in space and choose the numbers assigned so that the work done by the force when the particle moves from one point to another is simply the difference in the numbers assigned to those two points.

These numbers represent the potential energy of the particle. More precisely, the work done by the force as the particle moves from point A to point B is the potential energy at A minus the potential energy at B.

We can use the calculator to give a nice demonstration of the concept of potential energy associated with a conservative force.

<u>Problem 1.</u> The force $\vec{F} = 3x^2y^2\hat{i} + 2x^3y\hat{j}$ is conservative.

a. On a piece of graph paper lay out a coordinate grid extending (according to some convenient scale) from x=-1 m to x=+5 m and from y=-1 m to y=+5 m. Draw grid lines corresponding to 1 m intervals.

Take the potential energy to be 0 at the origin and calculate its value at each of the grid points. Do this by calculating the work done by the force as the particle moves from the origin to each of the grid points. The potential energy is then the negative of this work. It is best to work in a systematic way, going from one point to a neighboring point. Calculate the work done over 1 m segments and add to works previously calculated. At each grid point, write the value of the potential energy for that point.

When you are finished, you will have a chart showing the potential energy at the grid points for the given force. Denote this function by U(x,y).

b. Use the program discussed in section 8.4 to find the work done by $\vec{F}$ as the particle moves from the origin to the point x=3 m, y=5 m along each of the following paths.
   i.   the line x=0 from y=0 to y=5 m and then along the line y=5 m from x=0 to x=3 m.
   ii.   the straight line y=(5/3)x.
   iii. the curve y = 5sin(πx/6) where the angle is in radians.

These should all yield the same result and be equal to -U(3,5).

c. Use the program discussed in section 8.4 to find the work done by $\vec{F}$ as the particle moves from the point x=-1 m, y=+3 m to the point x=+5 m, y=0 along the straight line joining the two points. This should be equal to U(-1,3)-U(5,0).

Problem 2. The force $\vec{F} = (3x^2 - 6)y^2\hat{i} + 2x^3y\hat{j}$ is not conservative. Use the program discussed in section 8.4 to calculate the work done by this force as the particle moves from the origin to the point x=+3 m, y=+5 m along each of the following paths.

a. the line x=0 from y=0 to y=5 m and then along the line y=5 m from x=0 to x=3 m.
b. the straight line y=(5/3)x.
c. The curve y= 5sin(πx/6) where the angle is in radians.

Since the work done is different for different paths, you can see why it is impossible to devise a potential energy function as we did for the force of the preceding problem.

Chapter 9
ROTATIONAL MOTION

Rotational kinematics and dynamics for a rigid body rotating about a fixed axis are covered in this chapter. It supplements Chapters 11 and 12 of PHYSICS, Chapters 11 and 12 of FUNDAMENTALS OF PHYSICS, or similar chapters in other texts. There is a section on the conservation of angular momentum in which an inelastic "collision" between two rotating wheels is examined. Use of the programmable calculator allows us to study the interaction in detail and, given a torque law, to account for the rotational energy. It also allows us to consider the case when a net external torque is present. This section can be used for a study project. It supplements portions of Chapter 13 of PHYSICS, Chapter 12 of FUNDAMENTALS OF PHYSICS, and the sections on conservation of angular momentum in other texts.

9.1 Rotational Kinematics for Motion about a Fixed Axis

The purely kinematic quantities which are used to describe rotational motion about a fixed axis are the angular position θ, the angular velocity ω, and the angular acceleration α.

θ is the angle through which the body has turned. We measure it in radians and it is, in general, a function of time. We take the z axis to be the axis of rotation and take changes in θ to be positive if the thumb of the right hand points in the positive z direction when the fingers of the right hand curl around the axis in the direction of increasing θ.

The angular velocity ω is $d\theta/dt$ and is positive if θ increases with time. Conversely, it is negative if θ decreases with time. ω may be thought of as a vector, directed along the z axis. In this event, $\omega = d\theta/dt$ is the z component and may be either positive or negative, depending on the motion.

The angular acceleration α is $d\omega/dt$. This quantity may also be thought of as the z component of a vector $\vec{\alpha}$, which is also directed along the axis of rotation.

Angular variables are related to linear variables as follows. Put the origin of a coordinate system on the axis of rotation and let $\vec{r}$ be the position vector of a point in the rotating body. The velocity of that point is given by

$$\vec{v} = \vec{\omega} \times \vec{r} \qquad (9-1)$$

and its acceleration is given by

$$\vec{a} = \vec{\alpha} \times \vec{r} + \vec{\omega} \times \vec{v} . \qquad (9-2)$$

The point moves in a circle of radius R centered on the z axis. Its velocity has magnitude ωR. The first term in Eqn. 9-2 for the acceleration is tangent to the circle and this term is usually called the tangential acceleration. It is non-zero whenever the angular velocity ω of the body changes. That is, this term represents an acceleration which comes about because the body slows down or speeds up in its rotational motion. The second term in Eqn. 9-2 is the centripetal acceleration. It is a vector which points inward along the radius of the circle and it represents the change in the velocity which comes about because the direction of travel of the point is continuously changing as it moves around its circular orbit. It is non-zero even if $\alpha=0$.

Rotational kinematic problems for a fixed axis are mathematically very similar to one dimensional linear kinematic problems. θ replaces x, ω replaces v_x, and α replaces a_x in the equations and programs.

Given $\theta(t)$ as a function of time, it is possible to find the time t_f for the body to turn through a given angle θ_f. It is the root of the function $f(t)=\theta(t)-\theta_f$. If the motion is somewhat complicated, the body may have the same angular position more than once and there is then more than one root of $f(t)$.

The angular velocity ω as a function of time is found by differentiation of $\theta(t)$. This may be accomplished analytically in some cases or it may be accomplished by fitting $\theta(t)$ to a polynomial in t and differentiating the polynomial.

Similarly, the angular acceleration α as a function of time is found by differentiating ω. Once the root t_f of $f(t)$ is found, values of ω and α are computed by substituting t_f into the appropriate expressions for these quantities. The results give the angular velocity and angular acceleration when the body has reached the angular position θ_f.

The time for which the body is instantaneously at rest is a root of $\omega(t)$. Once this root has been found, the angular position and angular acceleration for that time can be found by direct substitution of the root into the expressions for $\theta(t)$ and $\alpha(t)$ respectively.

The root finding programs of Chapter 2 can be used to solve rotational kinematic problems just as they were used to solve linear kinematic problems. Exercises are provided in section 9.3.

9.2 Rotational Dynamics for Motion About a Fixed Axis

Tha angular momentum $\vec{\ell}$ about the origin of a particle at $\vec{r}$ is

$$\vec{\ell} = \vec{r} \times \vec{p} \tag{9-3}$$

where $\vec{p}$ is its momentum. For a system of particles, the total angular momentum $\vec{L}$ is the vector sum of the individual angular momenta. If the system is a rigid body rotating about the z axis, then the z component of the total angular momentum about the origin can be written

$$L_z = I\omega. \tag{9-4}$$

Here I is the moment of inertia of the body, given by

$$I = \sum m_i r_i^2 \tag{9-5}$$

where the sum is over all the particles which make up the body, m_i is the mass of particle i, and r_i is the distance from the axis of rotation to particle i.

The torque $\vec{\tau}$ about the origin, acting on a particle at $\vec{r}$, is given by

$$\vec{\tau} = \vec{r} \times \vec{F} \tag{9-6}$$

where $\vec{F}$ is the force acting on the particle. For a system of particles, we sum this quantity over all particles to find the total torque. For the systems we consider, the torque exerted by one particle, in the system, on another particle, also in the system, is equal in magnitude and opposite in direction to the torque exerted by the second particle on the first and these torques cancel each other in the sum. The total torque is exactly equal to the total external torque, the torque due to sources outside the system. We write for the total torque

$$\vec{\tau} = \sum \vec{r}_i \times \vec{F}_i \tag{9-7}$$

where the sum is over all external forces acting on the system. The force $\vec{F}_i$ acts at $\vec{r}_i$.

Torque causes the angular momentum to change according to

$$\vec{\tau} = d\vec{L}/dt \tag{9-8}$$

where $\vec{\tau}$ is the total external torque. For rotation of a rigid body about the z axis,

$$\tau_z = dL_z/dt = I\, d\omega/dt = I\alpha. \tag{9-9}$$

This equation is mathematically similar to Newton's second law for one dimensional motion. τ replaces the force, I replaces the mass, and α replaces the acceleration. If the torque is given as a function of time, the program of Fig. 6-3 can be used to integrate Eqn. 9-9 to obtain the angular position and angular velocity. Of course, the moment of inertia must also be given. The function f is τ_z/I.

Torque may also be a function of angular velocity and angular position. If it is a function of ω, the program of Fig. 6-1 can be used to integrate Eqn. 9-9.

In either case, the correspondence between physical quantities and the program storage locations is

        X1   angular position,
        X2   angular velocity,
        X3   time,
        X4   integration interval,
        X5   number of intervals,
and   X6   z component of torque divided by the moment of inertia.

An applied torque does work on the rotating body and, for rotation about a fixed axis, the work is given by the integral

$$W = \int_{\theta_0}^{\theta_f} \tau \, d\theta. \qquad (9\text{-}10)$$

If the torque is time or angular velocity dependent and the angular velocity is a known function of time, it is usually convenient to convert this integral to one with time as the variable of integration. Since $d\theta = \omega dt$, the result is

$$W = \int_{t_0}^{t_f} \tau(t,\theta,\omega)\omega \, dt. \qquad (9\text{-}11)$$

Time or angular velocity dependent torques are more common than torques which depend only on the angular position and Eqn. 9-11 provides the usual prescription for calculating the work done by an applied torque.

If τ is the total external torque acting on the body, then the work it does is equal to the change in the rotational kinetic energy:

$$\Delta K = \tfrac{1}{2}I\omega_f^2 - \tfrac{1}{2}I\omega_0^2 \qquad (9\text{-}12)$$

where ω_0 is the initial and ω_f is the final angular velocity.

The program of Fig. 8-1 can be used to evaluate Eqn. 9-11. The following storage assignments are made:

        X1   initial time,
        X2   final time,
and   X3   number of intervals.

The function which is evaluated at lines 130, 170, and 180 is $\tau\omega$.

9.3 Exercises

1. While a flywheel is brought up to speed, its angular position is given by

$$\theta(t) = 600t + 2\times10^4(e^{-0.03t} - 1).$$

Here θ is in radians and t is in seconds.

a. How much time does it take the wheel to make each of the first 3 revolutions ($\theta = 2\pi$, 4π, and 6π respectively)? It starts at t=0.

b. Show that
$$\omega(t) = 600(1 - e^{-0.03t})$$

and find the angular speed at the end of each of the first 3 revolutions.

c. Show that
$$\alpha(t) = 18e^{-0.03t}$$

and find the angular acceleration at the end of each of the first 3 revolutions.

d. For a point on the wheel 0.5 m from the axis, what is the speed, tangential acceleration, centripetal acceleration, and magnitude of the total acceleration at the end of each of the first 3 revolutions?

Notice in particular what happens to the tangential and centripetal parts of the total acceleration of the point as time goes on. As α decreases, so does the tangential component of the acceleration. At the same time, ω increases and so does the magnitude of the centripetal component of the acceleration. After a few minutes, the wheel reaches what is essentially a constant angular speed of 600 radians/s. Then the tangential component vanishes and the centripetal component becomes constant in magnitude.

2. The angular position of a flywheel being brought to rest is given by

$$\theta(t) = 100t - 2 \times 10^{-5} t^5 e^{0.01t}.$$

Here θ is in radians and t is in seconds. This equation describes the angular position until the wheel stops. Thereafter, it remains at rest.

a. Show that
$$\omega(t) = 100 - 1 \times 10^{-4} t^4 e^{0.01t} (1 + 2 \times 10^{-3} t)$$

and that the angular velocity is 100 radians/s at t=0.

b. Find the time it takes for the wheel to come to rest.

c. Find the angle through which the wheel rotates while it is brought to rest.

3. In some cases, frictional torque (in the bearing, for example) may vary linearly with the angular velocity. We consider a frictional torque given by an equation of the form

$$\tau = -A - B\omega \quad \text{for } \omega>0$$

and

$$\tau = +A - B\omega \quad \text{for } \omega<0.$$

A and B are positive constants. The equation changes with the sign of ω so that τ and ω are always in opposite directions.

a. By direct differentiation, show that, for $\omega>0$,

$$\omega(t) = -\frac{A}{B} + \left(\frac{A}{B} + \omega_0\right)e^{-Bt/I}$$

satisfies $\tau = I\, d\omega/dt$ and

$$\theta(t) = -\frac{A}{B}t + \frac{I}{B}\left(\frac{A}{B} + \omega_0\right)\left(1 - e^{-Bt/I}\right)$$

satisfies $\omega(t) = d\theta/dt$. Furthermore, show that $\omega(t)=\omega_0$ and $\theta(t)=0$ at $t=0$.

b. Consider a wheel with moment of inertia $I=3.2$ kg·m^2, initially rotating at 350 radians/s. If a torque of the above form, with A=5 N·m and B=0.06 N·m·s acts on the wheel, how much time elapses before the wheel stops? How many revolutions does the wheel make while it is being stopped?

c. An analytic expression for the stopping time can be found. Verify that $\omega=0$ for $t=t_s$, where

$$t_s = \frac{I}{B}\ln\left(1 + \frac{B}{A}\omega_0\right).$$

Evaluate this expression for the parameters given in part b and compare your answer with the answer you found using numerical techniques.

d. Use the program of Fig. 8-1 to calculate the work done by the torque during the stopping process. This is best done by using t as the variable of integration. Pick a suitable integration interval so that your answer is accurate to three significant figures. Compare your answer with $\frac{1}{2}I\omega_0^2$.

4. A wheel has moment of inertia $I=1.7$ kg·m^2 and initially rotates with an angular velocity of $\omega_0= 600$ radians/s. Starting at $t=0$, a torque with magnitude given by $0.5t^2$ (in SI units) acts to slow the wheel. Use either the program of Fig. 6-3

or the program of Fig. 8-1, with Δt=0.05 s, to answer the following questions.
a. After 5 s, what is the angular velocity of the wheel?
b. After 5 s, through what angle has the wheel turned?
c. How long does it take for the wheel to stop? (Use a polynomial fit.)
d. Through what angle does the wheel turn from t=0 to the time it stops?

5. A wheel with moment of inertia $I=2.3$ kg·m^2 is accelerated from rest to a final angular velocity of $\omega_f=350$ radians/s by the applied torque

$$\tau = 0.03(500 - \omega)^2.$$

Here τ is in N·m and ω is in radians/s. Obtain 3 significant figure answers.
a. How long does it take for the wheel's angular velocity to reach ω_f?
b. Through what angle does the wheel rotate as it is being accelerated?
c. How much work does the applied torque do as the wheel is accelerated? Use the work energy theorem.

9.4 Conservation of Angular Momentum

If the total external torque acting on the system is zero, then the total angular momentum is constant. The principle of conservation of angular momentum can be used to solve problems in many situations. Usually the system consists of two parts which interact with each other; each part exerts a torque on the other and the angular velocity of each part changes. Since there are no external torques acting, total angular momentum is conserved and we can use the equation which expresses its conservation to solve for one of the quantities which appear in the equation.

The conservation of angular momentum is sometimes demonstrated by considering the inelastic "collision" between two rotating wheels.

Problem 1. We consider two fly wheels which are mounted on the same axle but are free to turn independently of each other, at first. By sliding one wheel along the axle, the two wheels can be brought into contact with each other, face against face. The first wheel has moment of inertia $I_1=2.5$ kg·m^2 and is initially rotating with angular velocity $\omega_0=100$ radians/s. The second wheel has moment of inertia $I_2=1.5$ kg·m^2 and is initially at rest. The wheels are now brought into contact and eventually they rotate with the same angular velocity ω_f. During the collision, no torques act except those of each wheel on the other.

a. Use the conservation of angular momentum to show that

$$\omega_f = \frac{I_1 \omega_0}{I_1 + I_2} = 62.5 \text{ radians/s}.$$

b. If we know the torque of one wheel on the other, we can examine the collision process in detail. We suppose a torque given by

$$\tau = 8 + 2e^{-0.5t}$$

is exerted by wheel 1 on wheel 2 and a torque of the same magnitude but opposite sign is exerted by wheel 2 on wheel 1. Here τ is in N·m and t is in seconds.

Use either the program of Fig. 6-3 or the program of Fig. 8-1 to make a table giving the angular position and angular velocity of each of the wheels every 1 s from t=0, when they first touch, to t=13 s. Use an integration interval of $\Delta t=0.05$ s. The table extends to the non-physical situation for which wheel 2 spins faster than wheel 1. We need these points for part d.

c. For each of the times in the table, calculate the angular momentum of wheel 1, the angular momentum of wheel 2, and the sum.

Note that your results can be interpreted to show that, by means of the torques acting, angular momentum is transferred from one wheel to the other, while the total remains constant.

d. Fit $\omega_1 - \omega_2$ to a polynomial valid for times near its root and find the root. This is the time for which $\omega_1 = \omega_2 (=\omega_f)$. Evaluate ω_f and compare your answer with the answer to part a.

The nature of the torque influences the details of the interaction: the angular acceleration and hence the angular velocity and angular position of each wheel are different functions of time for different torques of interaction. However, as long as the torques acting on the two wheels are equal in magnitude and opposite in sign, angular momentum is conserved.

Problem 2. To test this assertion, we consider another form for the torque. We can imagine a spring-like force acting between the wheels such that the magnitude of the torque exerted on each wheel is proportional to the difference in their angular positions. The torque on wheel 1, due to wheel 2, is

$$\tau_1 = -k(\theta_1 - \theta_2)$$

and the torque on wheel 2, due to wheel 1, is

$$\tau_2 = +k(\theta_1 - \theta_2).$$

In each case, the torque tends to pull each wheel toward the other in angular position.

a. Take k=0.75 N·m/radian, use the program discussed in section 6.2, and calculate the angular position and angular velocity of each of the wheels for every 1 s between t=0 and t=10 s. The moments of inertia and the initial angular velocities are the same as in problem 1. At t=0, when the wheels start interacting, $\theta_1 = \theta_2 = 0$. Use an integration interval of $\Delta t = 0.05$ s.
b. Calculate the angular momentum of each wheel and the sum for each of the times of part a.

Examine the data produced. The wheels continue to rotate in the same general direction, but first one and then the other pulls ahead. The one with the larger θ pulls on the other, speeding it up while the one with the smaller θ slows the first. Angular momentum is transferred from wheel 1 to wheel 2 and back again, but through it all, the total angular momentum is conserved.

During the processes described in problems 1 and 2, the original kinetic energy of wheel 1 ($\frac{1}{2} I_1 \omega_0^2$) changes. Some is picked up by the second wheel and some is transformed to other forms of energy. The torques, of course, are the agency by which this transfer occurs.

Problem 3. Consider the situation given in problem 1. Use the program of Fig. 8-1 to calculate the work done by wheel 2 on wheel 1 during the first 10 s:

$$W_1 = \int_0^{10} \tau_1 \omega_1 \, dt$$

where τ_1 is the torque acting on wheel 1 and ω_1 is the angular velocity of wheel 1. Take the integration interval to be 1 s and use the data you generated in solving problem 1. Also calculate the work done by wheel 1 on wheel 2 during the same time:

$$W_2 = \int_0^{10} \tau_2 \omega_2 \, dt$$

where τ_2 is the torque acting on wheel 2 and ω_2 is the angular velocity of wheel 2.

Note that these works have opposite signs (wheel 1 slows down while wheel 2 speeds up), but they do not cancel. Compare the total work done with the change in the total kinetic energy. Kinetic energy is lost and it is lost by the agency of the work done by the torques.

For this problem, we do not know in detail where the energy goes. If the origin of the torques is frictional in nature, the energy goes to increase the internal energy of the wheels and they warm.

For problem 2, the mechanism of energy transfer is prescribed and we know the energy is stored as potential energy in the spring. In fact, as time goes on, kinetic energy is converted to potential energy and back again. The potential energy can be computed using

$$U = \tfrac{1}{2}k(\theta_1 - \theta_2)^2. \tag{9-13}$$

Problem 4. Consider the situation given in problem 2 and use the program of Fig. 8-1 to calculate the work done on wheel 1, the work done on wheel 2, and the total work done, all during the first 8 s. The first two of these are given by

$$W_1 = \int_0^8 \left[-k(\theta_1 - \theta_2)\omega_1 \right] dt$$

and
$$W_2 = \int_0^8 \left[+k(\theta_1 - \theta_2)\omega_2 \right] dt$$

respectively. Compare the total work done with the change in the total kinetic energy and to the change in the potential energy stored in the spring. The torques acting are the mechanism by which energy in converted from kinetic to potential form. Note the sum of the kinetic and potential energies is the same at t=8 s as at t=0.

For the total angular momentum to be conserved, it is important that the net external torque be zero. We can easily see the influence of an external torque.

<u>Problem 5.</u> Consider the situation of problem 1, but now suppose the torque acting on wheel 1 is $-10-2e^{-0.5t}$ while the torque acting on wheel 2 is $+8+2e^{-0.5t}$. Here the torques are in N·m and the time is in seconds. This represents a net external torque of -2 N·m.

Use the same initial conditions as given in problem 1, the program discussed in section 6.2, and find the angular momentum of each wheel and the sum for every 1 s from t=0 to t=9 s.

The net external torque is -2 N·m, a constant, and, since $\tau_{ext} = dL_{total}/dt$, the external torque should produce a change in the total angular momentum of

$$\Delta L = \tau_{ext} \Delta t = -2 \times 9 = -18 \text{ kg·m/s}$$

over the first 9 s. Do your results agree? Which wheel suffers the change (compared to the case of zero external torque) or is the change shared?

Chapter 10
PHYSICS OF SPECIAL SYSTEMS

Oscillatory motion, motion in a gravitational field, and rocket motion are considered. Numerical solutions to Newton's second law are obtained and the results are used to investigate interesting details of the motions. Damped and forced oscillatory motions are studied, Kepler's laws are demonstrated, and rockets in a gravitational field are examined. Each of the three sections are independent of the others and provide sufficient material for a separate study project. The three topics are covered in Chapters 9 (rockets), 15 (oscillations), and 16 (satellites) of PHYSICS and in Chapters 9 (rockets), 14 (oscillations), and 15 (satellites) of FUNDAMENTALS OF PHYSICS, as well as similar sections of other texts.

10.1 Oscillators

An ideal spring exerts a force on a mass which is proportional to the displacement of the mass from the equilibrium point (the point where the force is zero). If the mass is constrained to move along the x axis and if x=0 is the equilibrium point, then the force is given by

$$F = -kx \qquad (10\text{-}1)$$

where x is the position of the mass and k is a constant of proportionality, called the spring constant.

Substitution of the force law into Newton's second law yields a differential equation for the position x(t) of the mass as a function of time:

$$m\, d^2x/dt^2 = -kx. \qquad (10\text{-}2)$$

This differential equation has the general solution

$$x(t) = A\cos(\omega_0 t + \phi) \qquad (10\text{-}3)$$

where $\omega_0 = \sqrt{k/m}$. ω_0 is called the natural angular frequency of the oscillator and is measured in radians per second.

A and ϕ are constants of the motion, determined by the initial position and velocity of the mass. If the initial position is x_0 and the initial velocity is v_0, then

$$A = \left[x_0^2 + (v_0/\omega_0)^2\right]^{\frac{1}{2}} \tag{10-4}$$

and

$$\phi = \arctan\left(-\frac{v_0}{x_0\omega_0}\right). \tag{10-5}$$

Both equations are ambiguous in that either root can be used for A and there are two angles which have a given tangent. By convention, A is taken to be positive and ϕ is selected so that $A\cos\phi$ has the correct sign (matches the sign of x_0) or $-A\sin\phi$ has the correct sign (matches the sign of v_0). A is called the amplitude of the motion and it is the largest positive displacement of the mass during its motion. ϕ is called the phase constant of the motion.

The motion is periodic. It repeats itself in an interval of time T called the period. The period is given by $T = 2\pi/\omega_0$.

The kinetic energy of the mass is $K = \frac{1}{2}mv^2 = \frac{1}{2}m\omega_0^2 A^2 \sin^2(\omega_0 t + \phi)$, the potential energy is $U = \frac{1}{2}kx^2 = \frac{1}{2}kA^2\cos^2(\omega_0 t + \phi)$, and the total energy is $E = K + U = \frac{1}{2}m\omega_0^2 A^2$. Here $x=0$ is taken to be the zero of potential energy. It is noteworthy that the total energy is independent of the time and that it is proportional to the square of the amplitude.

Eqns. 10-1 and 10-2 may require modification for some situations. The mass may, for example, be subjected to a force of friction or of air resistance. The spring may not be ideal and, consequently, may exert a force which is different from $-kx$. If the spring is hung vertically, the force of gravity must be taken into account. Another situation of great practical importance occurs when a sinusoidally varying external force is applied to the mass.

All of these situations can be studied by numerically solving Newton's second law. It should be mentioned that many of these problems can also be solved analytically. However, we use numerical techniques here.

Problem 1. In order to study the motion of a mass under the influence of an ideal spring, consider a 2 kg mass attached to a spring with spring constant k=350N/m and released from rest at x=0.07 m.

a. Use the program of Fig. 6-3 to find the position and velocity of the mass at intervals of 0.05 s from t=0 (the time of release) to t=1 s. Take Δt=0.005 s.

b. Plot x(t) vs. t using the data generated in part a. Estimate the amplitude from the graph and compare the value with 0.07 m. Estimate the period from the graph and compare the value with $2\pi\sqrt{m/k}$. Compare ϕ with the value given by Eqn. 10-5.

c. For each point, calculate the kinetic energy, the potential energy, and the total energy. Compare the total energy to the original potential energy $\frac{1}{2}kx_0^2$.

At times when the spring is near its largest extension, the potential energy is large and the kinetic energy is small. When the mass is near x=0, the potential energy is small and the kinetic energy is large. Energy is transferred from kinetic to potential form and back again. The agent of this transfer, of course, is the work done by the force of the spring.

The curve plotted in part b, properly interpreted, represents all possible motions of the mass for the given total energy. Other motions are obtained if the initial position x_0 and initial velocity v_0 are different. If the combination $x_0^2+(v_0/\omega_0)^2$ is the same, then the energy and amplitude are the same and only the phase angle ϕ is different. The curve x vs. t is the same but it is shifted to the left or right relative to t=0.

d. On your graph, locate t=0.08 s and note the position and velocity of the mass for that time. We start a second, identical, oscillator so that its initial position and velocity are the same as the position and velocity, respectively, of the first oscillator for t=0.08 s.

First, use ϕ=arc tan$(-v_0/\omega_0 x_0)$ and $A=\left[x_0^2+(v_0/\omega_0)^2\right]^{\frac{1}{2}}$ to calculate ϕ and A for the second oscillator. Here x_0 and v_0 are the values you just read from the graph. ϕ should be different and A the same as for the first oscillator. Use the program of Fig. 6-3 to calculate x and v at t=0.02 s, 0.04 s, and 0.06 s for the second oscillator. Note that these values duplicate, to within tolerable error, the position and velocity of the first oscillator for t=0.1 s, 0.12 s, and 0.14 s respectively.

Problem 2. Suppose the oscillator of problem 1 is also subjected to a resistive force, proportional to the velocity. Then $m\, d^2x/dt^2 = -kx - bv$.

a. Take m=2 kg, k=350 N/m, b=2.8 kg/s, x_0=0.07 m, v_0=0, and use the program of Fig. 6-1, with Δt=0.002 s, to find x(t) and v(t) at intervals of 0.05 s from t=0 to t=1 s. Make a plot of x(t) vs. t. For later use, also find x(t) and v(t) at t=0.02 s, 0.04 s, 0.06 s, and 0.08 s. Use Δt=0.005 s for these points.

Notice the decrease in amplitude as times goes on. At the maximum points, the kinetic energy vanishes and the total energy is $\frac{1}{2}kx^2$. The decrease in amplitude is indicative of an energy loss. This is to be expected, of course, since the force of resistance does negative work.

b. For each of the times of part a, calculate the kinetic energy, the potential energy, and the total energy. Plot the total energy as a function of time.

c. We can verify the energy balance easily. Fit v^2 to a polynomial in t and integrate to find the work done by the resistive force from t=0 to t=0.08 s. Compare your result with the loss in total energy suffered by the mass in the same time interval.

d. Does the drag force also change the period of the motion? Define the period as the time interval from one maximum displacement to the next and estimate its value from the graph of part a. Compare the result with $2\pi\sqrt{m/k}$ and answer the question.

The oscillator studied in problem 2 is said to execute damped harmonic motion. In that case the motion is said to be underdamped since the mass continues to oscillate, albeit with ever decreasing amplitude. If the value of b is increased sufficiently, no oscillations occur and the motion is said to be overdamped.

Problem 3. Take m=2 kg, k=350 N/m, b=110 kg/s, x_0=0.07 m, v_0=0, and calculate x(t) for every 0.05 s from t=0 to t=1 s. Plot x(t) vs. t. Notice that no oscillations occur.

The motion of the mass can be changed considerably by an external force. We consider a force given by $F_0\cos(\omega t)$ where F_0 and ω are constants. The angular frequency ω of the impressed force is measured in radians per second and should not be confused with the natural frequency $\omega_0 = \sqrt{k/m}$ of the mass and spring. An impressed force such as the one given above can be generated, for example, by attaching the mass to one end of the spring and moving the other end back and forth in sinusoidal motion.

Newton's second law is $m\, d^2x/dt^2 = -kx + F_0\cos(\omega t)$. The resulting motion is the sum of two sinusoidal motions, one at the natural frequency and one at the impressed frequency. If these two frequencies are very different and F_0 is small, the motion at the natural frequency dominates.

Problem 4. To see this, consider the oscillator of problem 1 (m=2 kg, k=350 N/m, x_0=0.07 m, and v_0=0). The natural frequency is $\omega_0 = \sqrt{k/m} = \sqrt{350/2}$ = 13.2 radians/s. For each of the following cases, use the program of Fig. 6-3, with Δt=0.002 s, to plot x(t) every 0.05 s from t=0 to t=1 s.
a. F_0=18 N, ω=35 radians/s.
b. F_0=18 N, ω=15 radians/s.

For the graph of part a, the motion is nearly sinusoidal with a period of about 0.47 s (corresponding to a 13.2 radian/s angular frequency). The influence of the impressed force is seen in the slight deviations from a sinusoidal shape; in particular, in the slightly increased amplitude and the asymmetry of the function.

When ω is closer to ω_0, as it is in part b, the influence of the impressed force is more pronounced. The most important effect is the increased amplitude. As ω approaches closer to ω_0, the amplitude grows until, at $\omega=\omega_0$, the amplitude becomes infinite for an oscillator without damping.

The increase in amplitude can easily be accounted for in terms of the work done by the impressed force. If ω is different from ω_0, the impressed force is in the same direction as the velocity of the mass over some portions of the motion and in the opposite direction over other portions. The net work done over a time long compared with the period is almost zero. On the other hand, when $\omega \simeq \omega_0$, the impressed force and the velocity are in the same direction over a large portion of the motion. As you observe, the amplitude grows with time. For ω just slightly different from ω_0, the amplitude does not continue to grow but settles down eventually, so that the impressed force does positive and negative work in equal amounts and, again, no net work is done over times long compared with the period.

c. For each of the points calculated in part a, compute the total energy $E = \frac{1}{2}kx^2 + \frac{1}{2}mv^2$ and plot this quantity as a function of time. On the same graph paper, do the same for the points of part b. Identify the intervals of time when the impressed force is doing positive work and when it is doing negative work.

When the oscillator moves in a resistive medium, the natural motion, with angular frequency ω_0, is damped and the motion impressed by the external force dominates. The impressed force continues to supply the energy lost to the resistive medium and so the impressed motion is not damped. Newton's second law is now

$$m \, d^2x/dt^2 = -kx - b(dx/dt) + F_0 \cos(\omega t).$$

Problem 5. Consider the oscillator of problem 4 ($m=2$ kg, $k=350$ N/m, $F_0=18$ N, $\omega=35$ radians/s, $x_0=0.07$ m, and $v_0=0$) and take $b=15$ kg/s. Use the program of Fig. 6-1, with $\Delta t=0.002$ s, and plot $x(t)$ for every 0.04 s between $t=0$ and $t=1$ s.

We have considered a situation for which ω is considerably different from ω_0 and damping is fairly rapid. Notice that the motion near $t=0$ is similar to that for zero damping. The period is about 0.47 s and the natural motion dominates.

By $t=1$ s, the natural motion has been damped considerably and the resulting motion is that which is forced upon the mass by the impressed force. It is oscillatory with a period of about 0.18 s, which corresponds to an angular frequency of 35 radians/s.

10.2 Motion in a Gravitational Field

Any mass m_1 attracts any other mass m_2 with a gravitational force of magnitude

$$F = G \frac{m_1 m_2}{r^2} \tag{10-6}$$

where r is the distance between the masses and G is the universal gravitational constant (6.67×10^{-11} m^3/s^2 kg). The force of m_1 on m_2 is along the line joining the particles and is directed toward m_1. The force of m_2 on m_1 has the same magnitude and is oppositely directed.

For extended bodies, composed of many particles, the force is calculated as the sum of the forces all the particles in one body exert on all the particles in the other. The force may be a complicated function of the position and orientation of the bodies. If, however, the mass of each of the extended bodies is distributed with spherical symmetry, the force law can be written in the same form as for two particles, except that r is now the distance between the centers of the bodies.

For the problems of this section, we assume the force law, Eqn. 10-6, holds. Suns, moons, and planets are assumed to have spherical distributions of mass and spacecraft are assumed to be small compared with distances over which the gravitational force changes. They are considered to be particles.

We also assume, in each case, that one of the two bodies is much more massive than the other so that the center of mass of the system is very nearly at the center of the more massive body. We assume the more massive body is stationary with its center at the origin of the coordinate system. The second, less massive, body moves in a plane determined by its initial position vector and its initial velocity vector. We take this plane to be the x,y plane.

If x and y are the coordinates of the satellite (the less massive body), then the components of the force acting on it are given by

$$F_x = -G \frac{Mm}{r^3} x \qquad (10-7)$$

and
$$F_y = -G \frac{Mm}{r^3} y \qquad (10-8)$$

respectively. Here M is the mass of the central body and m is the mass of the satellite. Newton's second law produces

$$a_x = -G \frac{M}{r^3} x \qquad (10\text{-}9)$$

and
$$a_y = -G \frac{M}{r^3} y, \qquad (10\text{-}10)$$

once the mass of the satellite is cancelled from the two sides of each equation.

The problems of this section consist of using the program of Fig. 7-2, suitably modified, to integrate these equations for various initial conditions. There are a number of ways the program can be made more efficient for gravitational field problems. We note that the quantity $-GM/r^3$ appears in the equations for both the x and y components of the force and this quantity need be calculated only once each time the machine executes the instructions of the loop.

The flow chart for the revised program is shown in Fig. 10-1. The logic is exactly the same as for the program of Fig. 7-2. Storage assignments are

> X1   x coordinate,
> X2   y coordinate,
> X3   x component of velocity,
> X4   y component of velocity,
> X5   time,
> X6   integration interval,

and   X7   number of intervals before results are displayed.

Once the force is substituted into $\vec{v}_{new} = \vec{v}_{old} + (\vec{F}/m)\Delta t$, this equation becomes, in component form,

$$v_{xnew} = v_{xold} - \frac{GM}{r^3} x \, \Delta t \qquad (10\text{-}11)$$

and
$$v_{ynew} = v_{yold} - \frac{GM}{r^3} y \, \Delta t. \qquad (10\text{-}12)$$

The quantity GM appears in both the equations so it is computed separately at line 110 and stored in X10. The user must supply the mass of the central body. At lines 130, 180, and 250, GM is multiplied by Δt and divided by r^3, with the result stored

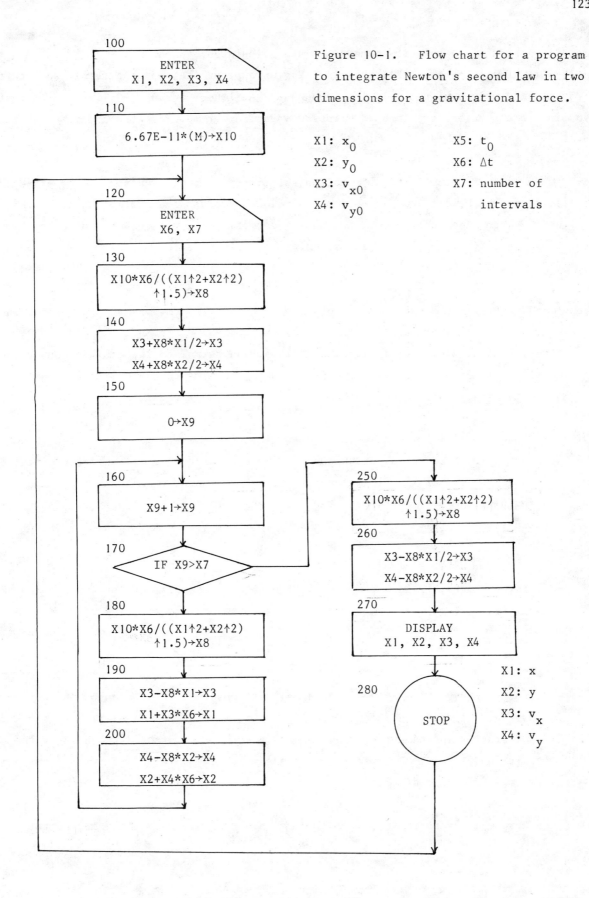

Figure 10-1. Flow chart for a program to integrate Newton's second law in two dimensions for a gravitational force.

X1: x_0 X5: t_0
X2: y_0 X6: Δt
X3: v_{x0} X7: number of
X4: v_{y0} intervals

X1: x
X2: y
X3: v_x
X4: v_y

in X8. When the new velocity components are calculated, the quantity in X8 is multiplied by either x (in X1) or y in (X2), as appropriate, and the result subtracted from the old values of the velocity component. At lines 140 and 260, we are interested in finding the velocity half a time interval earlier or later so the change is divided by 2.

In order to study satellite motion, we first use the calculator to generate an orbit, then use the data to answer some questions or test some statements about the motion. We can also obtain some important information about satellite motion by changing the initial conditions and calculating the new orbits.

<u>Problem 1.</u> Consider a satellite in orbit about the earth ($M=5.98\times10^{24}$ kg). At t=0 it is at $x=7.2\times10^6$ m, y=0 and has velocity $v_x=0$, $v_y=9\times10^3$ m/s.

a. Use the program of Fig. 10-1, with $\Delta t=20$ s, and calculate the position and velocity of the satellite for every 500 s from t=0 to the time the satellite has completed a little more than one revolution. Plot the orbit, y vs. x.

To answer the following questions you will need to make some calculations using the data generated in answer to part a. If the storage capacity of your machine is large enough, it is well worthwhile to modify the program so that the energy and angular momentum are calculated for each of the points of part a. The instructions are given below. They should be inserted after line 260 in the program.

b. The force of gravity is a conservative force. A potential energy function can be defined. If the potential energy is taken to be zero at large r, then

$$U = -GMm/r.$$

The total energy (kinetic plus potential) is conserved. For each of the points of part a, calculate the total energy per unit mass of the satellite. This number is not exactly the same for every point since there is error introduced by the integration process. However, it is the same to 5 significant figures. Furthermore, the error can be reduced by using a smaller value for the interval.

c. Since the force is a central force (it is parallel to the position vector $\vec{r}$), the torque it exerts on the satellite is zero ($\vec{r}\times\vec{F}=0$) and the angular momentum of the satellite is conserved. To verify this, calculate the z component of the angular momentum per unit mass for each of the points of part a. Since the orbit lies in the x,y plane, the other components automatically vanish.

To answer parts b and c, we make use of the equations

$$E/m = \tfrac{1}{2}(v_x^2 + v_y^2) - GM/(x^2 + y^2)^{\tfrac{1}{2}} \tag{10-13}$$

and
$$L_z/m = xv_y - yv_x. \tag{10-14}$$

The appropriate statements to follow line 260 are

.5*(X3↑2+X4↑2)-X10/SQRT(X1↑2+X2↑2)→X11

and   X1*X4-X2*X3→X12

Then X11 contains the energy and X12 contains the z component of the angular momentum, both per unit mass of the satellite. These numbers must be displayed.

One of the consequences of the conservation of angular momentum is that the radius vector, from the origin to the satellite, sweeps out equal areas in equal time intervals as the satellite moves along its orbit. This is one of Kepler's laws of planetary motion.

The points you have plotted give the position of the satellite at equal time intervals. All that remains to verify the law of equal areas is to calculate the area enclosed by two vectors drawn to two successive points on the orbit. The two points are connected by the curved orbit, but we approximate that portion of the orbit by a straight line joining the points.

Suppose x_1,y_1 and x_2,y_2 represent two successive positions of the satellite, at times which differ by 300 s. The corresponding radius vectors are $\vec{r}_1 = x_1\hat{i} + y_1\hat{j}$ and $\vec{r}_2 = x_2\hat{i} + y_2\hat{j}$. The area of the triangle defined by the origin and these two points is

$$A = \tfrac{1}{2}|\vec{r}_1 \times \vec{r}_2| = \tfrac{1}{2}|x_1 y_2 - x_2 y_1|. \tag{10-15}$$

Problem 2. For the satellite of problem 1, calculate the area A for two successive points near the starting point, for two successive points near the place where x has its greatest negative value and for two successive points near the place where y has its greatest positive value. Notice that the areas are equal to within computational error. Notice also that the original plot you made to answer part a of problem 1 is highly suggestive of the law of areas. At places where the satellite is near the origin, it moves a fairly great distance in 300 s, whereas at places where the satellite is far from the origin, it does not move nearly so far in the same time.

The orbit is an ellipse. This is another of Kepler's laws of planetary motion. It is easy to verify, as we see in the next problem.

Problem 3. Recall the definition of an ellipse. It is a figure such that the sum of the distances from any point on the figure to two fixed points, called focii, is the same as for any other point on the figure.

The first part of the problem is to locate the focii. For the orbit of problem 1, one of the focii is at the origin, the other is on the x axis, the same distance from the center of the figure as is the first focus. The two focii are on opposite sides of the figure center.

Find the geometric center of the figure. It is on the x axis, halfway between the two points where the orbit crosses that axis. One of these points is at $x=7.2\times 10^6$ m. Find the other by fitting $x(t)$ and $y(t)$ to polynomials with $t=7.5\times 10^3$ s as the central point. Find the root of $y(t)$ and substitute it into the polynomial for x. Suppose the second crossing is at $x=x_1$, a negative number. Then the geometric center of the orbit has x coordinate $x=(x_1+7.2\times 10^6)/2$ and the second focus has x coordinate $x=x_1+7.2\times 10^6$ m.

Let x_2 be the x coordinate of the second focus and let $\vec{r}=x\hat{i}+y\hat{j}$ be the vector from the origin to any selected point on the orbit. The distance from the first focus to the point is $(x^2+y^2)^{\frac{1}{2}}$ and the distance from the second focus to the point is $|\vec{r}-x_2\hat{i}| = \left[(x-x_2)^2 + y^2\right]^{\frac{1}{2}}$. If the orbit is an ellipse, the quantity

$$s = (x^2 + y^2)^{\frac{1}{2}} + \left[(x-x_2)^2 + y^2\right]^{\frac{1}{2}}$$

is the same for all points x,y on the orbit.

Evaluate s for the two points on the orbit with y=0, for a point near the place where the orbit has its largest y coordinate, and for a point near x=0. If you obtain the same value of s for all these points, you should have some confidence that the orbit is an ellipse.

The third of Kepler's laws for planetary motion is: the square of the period (time for one revolution) is proportional to the cube of the semimajor axis. The exact relationship can be derived using Newton's second law. It is

$$T^2 = \frac{4\pi^2}{GM} a^3 ,$$

where a is the length of the semimajor axis.

The period is easy to find using data generated in problem 3. It is twice the time for the satellite to go from the starting point around to the point where y=0 again. Fit y(t) to a polynomial in the neighborhood of the place where the orbit crosses the x axis and find its root. The period is twice the root. The root was found in answer to problem 3. The semimajor axis has length equal to half the distance between the two points where the orbit crosses the x axis. These points were also found in answer to problem 3.

Problem 4. Find the period and length of the semimajor axis for the satellite of problem 1. Use the results of problem 3. Verify the above equation explicitly by comparing T^2 with $(4\pi^2/GM)a^3$.

For the following problem it is convenient to think of a space craft in the vicinity of a planet rather than a satellite in orbit. We consider motion in a gravitational field when the moving body is not bound to the massive stationary body and the orbit is not a closed figure.

If the total energy is zero or positive (still using large r as the zero of potential energy), the space craft moves in an orbit which is not an ellipse. It is instead a parabola (if E=0) or, as is more likely, a hyperbola (if E>0). The only quantity one needs to know to predict escape is the total energy. If E<0, the space craft is bound in an elliptic orbit; if E>0, it escapes along a hyperbolic trajectory.

<u>Problem 5</u>. Start the space craft in the gravitational field of the earth, at $x=7.2\times 10^6$ m, y=0 with velocity $v_x=0$, $v_y=1.2\times 10^4$ m/s.
a. Use the program of Fig. 10-1, with $\Delta t=20$s, and calculate the position and velocity for every 500 s from t=0 to t=4500 s. Make a graph of y vs. x and measure the angle at which the trajectory recedes to large r.
b. Calculate the total energy per unit mass from the initial conditions and verify that it is positive.

10.3 <u>Rocket Motion</u>

A rocket is propelled forward by the rearward ejection of mass (the fuel). We assume that during the time Δt, the rocket emits mass $-\Delta M$ at velocity $\vec{u}$ relative to the rocket and that, during this time, the velocity of the rocket changes from $\vec{v}$ to $\vec{v}+\Delta\vec{v}$. ΔM is the change in the mass of the rocket, a negative number, so $-\Delta M$ is positive. The velocity of the emitted mass is $\vec{u}+\vec{v}$ relative to the inertial frame used to measure the velocity of the rocket.

The momentum before the emission of the mass is $(M-\Delta M)\vec{v}$ and the momentum afterward is $M(\vec{v}+\Delta\vec{v})-\Delta M(\vec{u}+\vec{v})$. The first term is the momentum of the rocket and the second is the momentum of the ejected mass. The system consists of the rocket and expended fuel and we add the momenta of the two parts. The time rate of change of momentum is

$$\frac{d\vec{p}}{dt} = \lim_{\Delta t \to 0} \frac{M(\vec{v}+\Delta\vec{v}) - \Delta M(\vec{u}+\vec{v}) - (M-\Delta M)\vec{v}}{\Delta t}$$

$$= M\frac{d\vec{v}}{dt} - \frac{dM}{dt}\vec{u}. \qquad (10\text{-}16)$$

According to Newton's second law, this must be equal to the total external force acting on the system. So

$$\vec{F} = M\frac{d\vec{v}}{dt} - \frac{dM}{dt}\vec{u}. \qquad (10\text{-}17)$$

In the limit as Δt and ΔM become infinitesimal, the force on the ejected mass also becomes infinitesimal, and we take $\vec{F}$ to the external force on the rocket alone.

For example, in the gravitational field of a spherical planet of mass M_p, the rocket equation becomes

$$M \frac{d\vec{v}}{dt} = \vec{u} \frac{dM}{dt} - \frac{GM_p M}{r^3} \vec{r} \qquad (10\text{-}18)$$

where $\vec{r}$ is the position vector of the rocket relative to the center of the planet. It is the mass of the rocket and unexpended fuel which is used to calculate the gravitational force.

In writing Eqn. 10-18, we have transposed the term $\vec{u}\, dM/dt$ to the right side of the equation. This quantity is called the thrust of the rocket engine and it is responsible for driving the rocket. It depends on the rate at which the engine ejects mass as well as on the velocity of the ejected mass relative to the rocket.

Problem 1. For some situations, there exist analytic solutions to the rocket equation. We consider one: a rocket which expends fuel at a linear rate, so that $M = M_0 - Kt$ where M_0 is the initial mass of rocket and fuel and K is a constant. We suppose further that the rocket is fired in a uniform gravitational field for which the acceleration due to gravity has magnitude g. The rocket is fired upward in the positive y direction and the force of gravity is downward. Then

$$(M_0 - Kt) \frac{dv}{dt} = uK - (M_0 - Kt)g.$$

Here we have used $\vec{u} = -u\hat{j}$ and $\vec{v} = +v\hat{j}$. The positive y axis is upward.

This differential equation has the general solution

$$v(t) = C - u \ln(M_0 - Kt) - gt.$$

The constant C can be evaluated by applying the condition that $v = v_0$ at $t = 0$. When this is done, the equation

$$v(t) = v_0 + u \ln \frac{M_0}{M_0 - Kt} - gt$$

results.

a. Verify by direct differentiation and substitution that this equation is a solution to the rocket differential equation, Eqn. 10-17.

b. Consider a rocket which carries 80% of its original mass as fuel and burns it linearly for 5 s, at which time the fuel is gone and the engine shuts off. Assume the rocket starts from rest, take u=5000 m/s, and evaluate the above equation to find the velocity of the rocket at burn out.

c. Use the program of Fig. 6-3 to find the position and velocity of the rocket for every half second from t=0 to burn out. Since, for this rocket, $K=0.16M_0$, the rocket equation can be recast into the form

$$\frac{dv}{dt} = \frac{0.16u}{1-0.16t} - g.$$

It is the right side of this equation, with numerical values substituted for u and g, which should be used for the function f in the program. Use $\Delta t=0.01$ s for the integration interval.

Problem 2. The rocket described above in problem 1 is again fired in the earth's gravitational field but now we take into account the decrease in gravitational force with increase in the rocket's distance from the center of the earth. For this case

$$\frac{dv}{dt} = \frac{uK}{M_0-Kt} - \frac{GM_e}{y^2}.$$

Here M_e is the mass of the earth and y is the distance from the center of the earth to the rocket. This equation cannot be solved analytically.

Take $y_0=6.37\times10^6$ m, $v_0=0$, and numerically find the position and velocity every 0.5 s from t=0 to burn out. Use $\Delta t=0.01$ s. Also calculate the total energy per unit mass at burn out and tell whether or not the rocket is bound to the earth.

The next problem deals with a rocket which moves in two dimensions. It is a projectile problem but the projectile has thrust.

Problem 3. Suppose a model rocket is fired at an angle θ above the horizontal in a unifrom gravitational field. We neglect air resistance. If the x,y plane is the plane of the trajectory with the y axis in the upward direction, then

$$\frac{d\vec{v}}{dt} = -\frac{K\vec{u}}{M_0-Kt} - g\hat{j}.$$

Now $\vec{u}$ is always opposite to $\vec{v}$ and we may write

$$\frac{d\vec{v}}{dt} = +\frac{Ku}{M_0-Kt}\frac{\vec{v}}{v} - g\hat{j}$$

where u is the magnitude of $\vec{u}$ and we have used the unit vector $\vec{v}/v$ to describe the direction of $\vec{u}$. This equation holds while fuel is being ejected. After the burn time has elapsed, the rocket becomes an ordinary projectile and $d\vec{v}/dt=-g\hat{j}$.

Use the program of Fig. 7-1, which integrates Newton's second law for velocity dependent forces. If the rocket starts from rest, the machine will have trouble with $\vec{v}/v$. The sole purpose of this factor is to describe the direction of $\vec{u}$, so pick any small initial velocity in the correct direction.

Consider a small rocket with the following characteristics: its mass without fuel is 9×10^{-2} kg, the mass of the fuel is 1.3×10^{-2} kg, and the fuel ejection speed is 7.5×10^{-2} m/s. Fuel is ejected for 1.2 s and then has been completely expended.

Assume the rocket is fired from rest at an angle of 45° above the horizontal and find the range over level ground. Use Δt=0.005 s. The integration must be carried out in two steps since the acceleration changes at t=1.2 s. Do the second integration analytically, using the results of the first as initial conditions. Solve algebraically for the range.

Chapter 11

THE ELECTRIC FIELD

Programs to calculate the electric field due to a set of discrete charges and due to a continuous distribution of charge along a line are presented. These programs are used to make graphical representations of the field for various charge distributions and to verify qualitative statements about the magnitude, direction, and symmetry of the field. The field of a charge distribution as a superposition of the fields of individual charges is stressed. The material supplements Chapter 27 of PHYSICS, Chapter 24 of FUNDAMENTALS OF PHYSICS, and discussions of the electric field found in other texts.

11.1 The Electric Field of Point Charges

A point charge q, located at $\vec{r}\,'$, produces an electric field at all points in space. In particular, the field it produces at the point $\vec{r}$ is given by

$$\vec{E}(\vec{r}) = \frac{1}{4\pi\varepsilon_0} \frac{q}{|\vec{r}-\vec{r}\,'|^3} (\vec{r}-\vec{r}\,'). \qquad (11\text{-}1)$$

Here $1/4\pi\varepsilon_0$ is a constant whose value in SI units is 8.98755×10^9 N·m^2/C^2. The combination $\vec{r}-\vec{r}\,'$ is the displacement vector from the charge to the field point (the observation point or place for which the field is calculated). The magnitude of this vector is the distance from the charge to the field point and the magnitude of $\vec{E}$ varies as the reciprocal of the square of this distance. When the magnitude is calculated, $|\vec{r}-\vec{r}\,'|$ appears in the numerator and cancels one of the factors in the denominator to produce the $1/|\vec{r}-\vec{r}\,'|^2$ behavior.

The factor $(\vec{r}-\vec{r}\,')$ is needed in the numerator to describe the direction of the electric field. The electric field vector lies along the line joining the charge and the field point. If q is positive, the field points away from the charge; if q is negative, the field points toward the charge.

If there is more than one charge, the electric field at a point is the vector sum of the individual fields due to the various charges, each determined according to Eqn. 11-1. We designate the charges by subscripts: q_1, q_2, ... and, in general, charge i by q_i. If $\vec{r}_i$ is the position of charge i, then the electric field at $\vec{r}$ is given by

$$\vec{E}(\vec{r}) = \sum \frac{1}{4\pi\epsilon_0} \frac{q_i\,(\vec{r}-\vec{r}_i)}{|\vec{r}-\vec{r}_i|^3} \qquad (11\text{-}2)$$

where the sum is over the charges.

Some care must be exercised in evaluating Eqn. 11-2 since, near any one of the charges, the field is large and the calculator may not be able to handle so large a number. We evaluate $|\vec{r}-\vec{r}_i|$ first and check its value to make sure it is not too small. If it is, we ask the machine to tell us it cannot complete the calculation for the point $\vec{r}$ selected.

Fig. 11-1 shows the flow chart for the calculation of the electric field, given the charges and their positions. At line 100, the coordinates x,y,z of the field point are entered and stored in X1, X2, and X3 respectively. The number of charges is placed in X4. Storage locations for the components of the electric field are: E_x in X5, E_y in X6, and E_z in X7. At line 110, these quantities are set equal to zero in preparation for entering a loop where the contributions of the various charges are calculated and summed.

X8 counts the charges and it is also zeroed. At line 130, it is checked to see if all charges have been considered and, if they have, the machine displays the field components and goes to line 100, where coordinates of another field point are entered.

At line 140, a charge and its coordinates are entered and stored. X9 contains the charge (with appropriate sign), X10, X11, and X12 contain the x, y, and z components, respectively, of its position vector.

At 150, the quantity $|\vec{r}-\vec{r}_i|^2$ is calculated and stored in X13. Its magnitude is then compared with 1×10^{-60}. Any other convenient small number will do. If the square of the distance is too small, the calculation is not continued but, instead, the number 1×10^{-60} is displayed to signal the user that $\vec{r}$ is too close to the charge. If the machine is restarted after STOP is executed, it returns to line 100 to accept the coordinates of another field point.

If $|\vec{r}-\vec{r}_i|$ is not too small, the program continues at line 170 where the

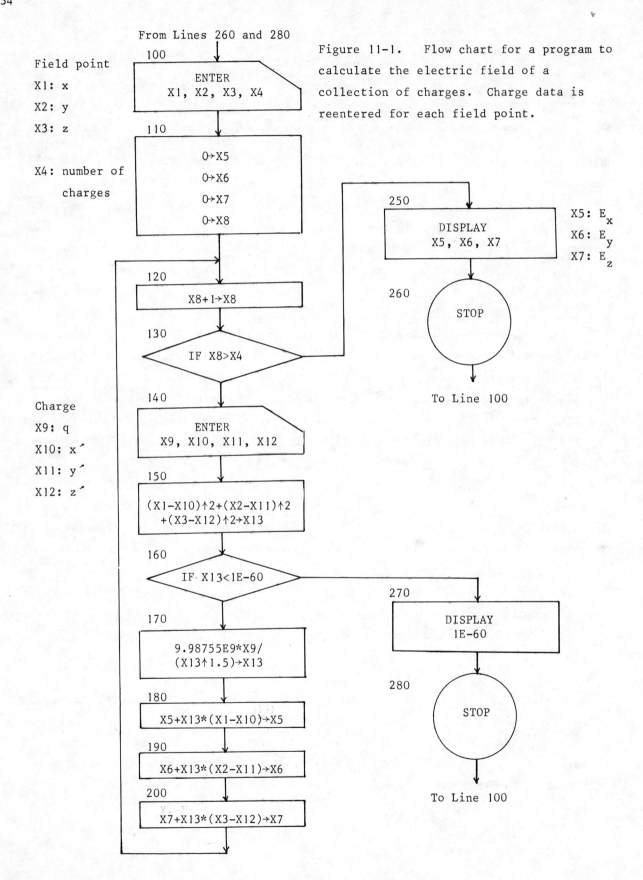

Figure 11-1. Flow chart for a program to calculate the electric field of a collection of charges. Charge data is reentered for each field point.

combination $(1/4\pi\epsilon_0)q_i/|\vec{r}-\vec{r}_i|^3$ is calculated. This combination occurs as a common factor in the equations for all three components of the field. It is more efficient to compute it once and store it then to compute it three times, once for each component.

At lines 180, 190, and 200, the x, y, and z components, respectively, of the electric field due to the charge being considered are calculated. They are added to components previously calculated to produce the components of the total field due to all charges considered so far. The machine then returns to line 120 where data for the next charge are entered.

This program uses a fairly small number of storage locations for the calculation of a vector quantity. It is, however, somewhat tedious to apply if the electric field is desired at more than a few field points. The complete set of charges and their positions must be entered for each field point. It is worthwhile to use a program for which the charges and their positions are entered and stored once, then recalled from storage as needed.

In order to write such a program, we introduce a new procedure, called indirect addressing, for naming storage locations. We allow the number which identifies the storage location to be a variable whose value is computed by the machine. For example, we write X(X1) to mean the storage location whose identification number is the number stored in X1. If the number 1 is in X1, then X(X1) is X1; if the number 2 is in X1, then X(X1) is X2 etc. To use this notation, we must be sure X1 contains an integer. The similarity between this notation and that for a function is evident.

We must carry the notation one more step. The quantity in the parentheses may be an arithmetic expression. For example X(3*X1+5) is a valid label for a storage location. The number in X1 (an integer) is multiplied by 3 and the result added to 5 to produce the number of the storage location sought.

We now return to the description of a program which stores the charges and their positions, then uses the stored data to calculate the electric field.

To reduce the number of storage locations required, we consider charges which lie in the x,y plane and calculate the x and y components of the field at a point in the x,y plane. For each charge we must store three numbers: the charge and two coordinates. The numbers are stored in the order q_1, x_1, y_1, q_2, x_2, y_2, q_3, x_3, ..., starting at X8. The first seven locations are needed for other parts of the calculation and we wish the charge information to be in the highest numbered locations so that the number of charges can be left an arbitrary number.

A numbering scheme which accomplishes this goal is as follows. If X4 is the number of the charge (the subscript in q_i), then we store the charge in location number 3*X4+5, the x coordinate of its position in location number 3*X4+6, and the y coordinate of its position in location number 3*X4+7. The following table should convince you that the scheme does the job. The content of each location is given in parentheses.

CHARGE NUMBER	X4	X(3*X4+5)	X(3*X4+6)	X(3*X4+7)
1	1	X8 (q_1)	X9 (x_1)	X10 (y_1)
2	2	X11 (q_2)	X12 (x_2)	X13 (y_2)
3	3	X14 (q_3)	X15 (x_3)	X16 (y_3)

Fig. 11-2 shows a flow chart for the program. It divides nicely into two parts, shown on separate pages because of the size of the program. In the first part, which consists of lines 100 through 140, charge information is entered and stored. In the second part, the field is calculated.

X1, entered at line 100, is the number of charges to be considered. X4 counts the charges as information about them is entered. At 110 it is zeroed in preparation for entering the loop which begins at line 120. There, X4 is incremented by 1. Then, at line 130, the charge, the x coordinate of its position, and the y coordinate of its position are entered and stored in X(3*X4+5), X(3*X4+6), and X(3*X4+7) respectively. If X4<X1, there are more charges to be considered and the machine returns to line 120.

When information about the last charge has been entered, X4=X1 and, at line 140, the transfer statement causes the machine to go to line 150, where calculation of the electric field begins.

X2 and X3 contain the x and y coordinates, respectively, of the field point. They are entered at line 150.

In the loop beginning at 170, the individual fields of the charges are calculated

Figure 11-2. Flow chart for a program to calculate the electric field of a collection of charges. Charge information is stored.

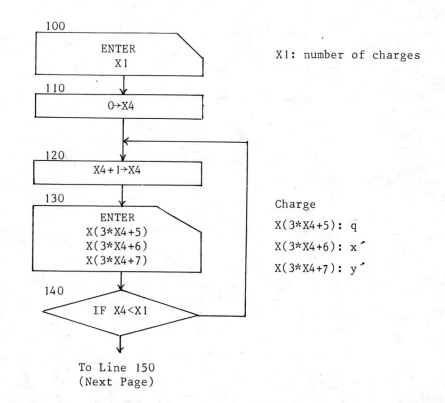

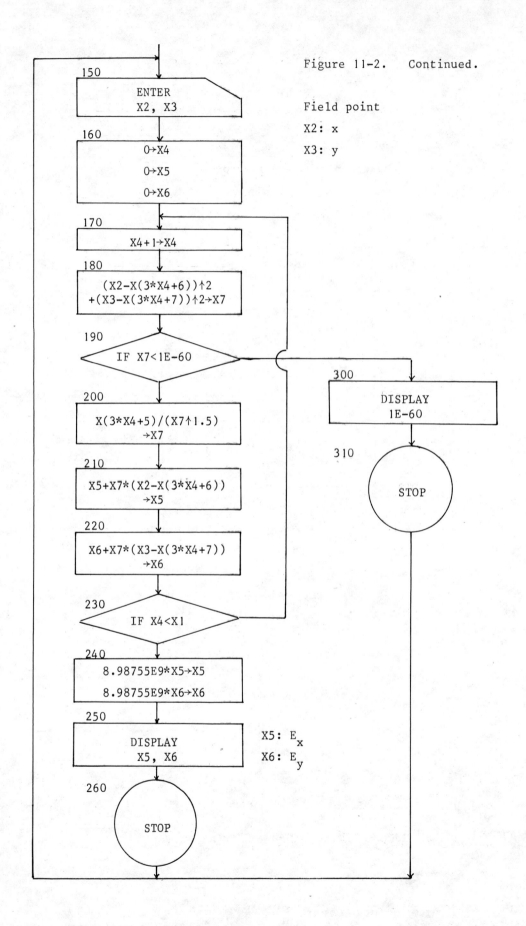

Figure 11-2. Continued.

Field point
X2: x
X3: y

X5: E_x
X6: E_y

one at a time and summed. X4 again counts the charges and must be zeroed before the loop is entered. X5 and X6 contain the x and y components, respectively, of the field (excluding the factor $1/4\pi\epsilon_0$) and these must also be zeroed prior to entering the loop. This is done at line 160.

At line 170, X4 is incremented by 1 each time the field of a new charge is calculated. Then, at 180, the quantity $|\vec{r}-\vec{r}_i|^2$ is calculated and, at 190, compared to 1×10^{-60} to see if the field point is too close to the charge. If it is, the number 1×10^{-60} is displayed and the machine stops. On restarting, it proceeds to line 150 to accept a new field point.

If $|\vec{r}-\vec{r}_i|^2$ is larger than 1×10^{-60}, the machine calculates $q_i/|\vec{r}-\vec{r}_i|^3$ and stores it in X7. The factor $1/4\pi\epsilon_0$ is not used at this time. Since it is a common factor for the fields of all charges, it is efficient to wait until the contributions of all charges have been summed before multiplying by it.

At 210 and 220, the x and y components, respectively, of the field (except for the $1/4\pi\epsilon_0$ factor) are calculated and added to previously computed components. If there is more charge to be considered, X4<X1, and the machine goes back to 170 to consider the next charge.

When all charge has been considered, the machine goes to line 240, where multiplication of each component by $1/4\pi\epsilon_0$ is carried out. The field components are displayed at line 250 and, on restarting, the machine proceeds to line 150 where the coordinates of a new field point are entered.

If the calculator has the ability to store programs and data on magnetic cards and then read them at appropriate times, then electric fields of large numbers of charges can be calculated on a hand-held machine. Otherwise, the number of charges considered is limited by the size of the memory of the machine.

11.2 Continuous Distributions of Charge Along a Line

In many situations of interest, charge is distributed along a line, over a surface, or in a volume so densely that, for purposes of calculation, it may be considered a continuous distribution. We consider distributions which are finite in extent. All of the charge is confined to a finite region of space and we find the electric field at points outside the distribution.

The simplest distribution to deal with is a line of charge. The distribution is characterized by a charge density which, for our purposes, is considered to be a function of position within the distribution. The line charge density, denoted by $\lambda(\vec{r}')$, gives the charge per unit length at the place $\vec{r}'$ (a point on the line). The density takes into account the sign of the charge. It is positive at places where there is positive charge and negative at places where there is negative charge.

We consider a line of charge on the x axis, from x=0 to x=a. A segment of length dx' at $\vec{r}'=x'\hat{i}$ contains an amount of charge equal to $\lambda(\vec{r}')dx'$ and produces an electric field at $\vec{r}$ which is given by

$$d\vec{E}(\vec{r}) = \frac{1}{4\pi\epsilon_0} \frac{\lambda(\vec{r}')\ (\vec{r}-\vec{r}')\ dx'}{|\vec{r}-\vec{r}'|^3}. \tag{11-3}$$

We use primed variables to describe the position of the charge in order to distinguish them from the unprimed variables used to describe the position of the field point.

To calculate the total field at $\vec{r}$ we sum the contributions of the various segments:

$$\vec{E}(\vec{r}) = \frac{1}{4\pi\epsilon_0} \int_0^a \frac{\lambda(\vec{r}')\ (\vec{r}-\vec{r}')\ dx'}{|\vec{r}-\vec{r}'|^3}. \tag{11-4}$$

If the field point is in the x,y plane, then the field has only x and y components, $\vec{r}=x\hat{i}+x\hat{j}$,

$$E_x(\vec{r}) = \frac{1}{4\pi\epsilon_0} \int_0^a \frac{\lambda(x')\ (x-x')\ dx'}{\left[(x-x')^2 + y^2\right]^{3/2}}, \tag{11-5}$$

and

$$E_y(\vec{r}) = \frac{y}{4\pi\epsilon_0} \int_0^a \frac{\lambda(x')\ dx'}{\left[(x-x')^2 + y^2\right]^{3/2}}. \tag{11-6}$$

The integration program of Fig. 8-1 can easily be modified to evaluate these integrals. When E_x is calculated, the integrand is $\lambda(x')(x-x')/\left[(x-x')^2+y^2\right]^{3/2}$, and when E_y is calculated, the integrand is $\lambda(x')/\left[(x-x')^2+y^2\right]^{3/2}$. The result for E_x must be multiplied by $1/4\pi\epsilon_0$ and the result for E_y must be multiplied by $y/4\pi\epsilon_0$. These multiplications can be programmed on the machine or can be carried out as separate steps after the integration program has been run.

We must enter the coordinates of the field point. Put x into X11 and y into X12 and augment the ENTER statement at line 100 to read

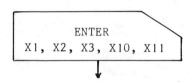

X1 is 0 and X2 is the length of the line of charge. X3 is still the number of intervals to be used in the integration.

The function to be evaluated is now a function of x', x, and y and the statement at line 130 should be f(X1,X10,X11)→X5, while the statements at lines 170 and 180 should be f(X1,X10,X11)→X9. For example, if $\lambda(x')=.5(x')^2$ and E_x is to be computed, the last statement would read

$$.5*(X1\uparrow 2)*(X10-X1)/(((X10-X1)\uparrow 2+X11\uparrow 2)\uparrow 1.5)\to X9$$

Following the STOP statemtnt at line 220, the machine should be instructed to return to line 100 to consider the other component or another field point.

The program described above embodies the basic technique for calculating the electric fields of continuous charge distributions. The distribution is broken into small elements, the contributuion to the field of each element is calculated, and the contributions of all the elements are summed.

There is however, a slightly different technique which, in many cases, requires fewer segments and shorter running time to achieve the same accuracy as the technique presented above.

We consider a situation for which λ is a slowly varying function of x'. For this case, it is the desire to reduce the variation of $1/|\vec{r}-\vec{r}'|^3$ rather than the variation of λ in a segment which leads us to use a large number of small segments. We again divide the line into N segments and the length of a segment is chosen so that λ can be considered constant within the segment. The exact expression for the electric field, given by Eqn. 11-4, can be written as a sum of integrals, one for each segment:

$$\vec{E}(\vec{r}) = \frac{1}{4\pi\varepsilon_0} \sum_{i=1}^{N} \lambda(x_i + .5a/N) \int_{x_i}^{x_{i+1}} \frac{(\vec{r}-\vec{r}')\,dx'}{|\vec{r}-\vec{r}'|^3} . \qquad (11-7)$$

Here segment i begins at x_i and ends at x_{i+1}. Its center is at $x_i + .5a/N$ since $x_{i+1} - x_i = a/N$. It is assumed that λ is constant in the segment and it is evaluated at the center of the segment.

The usefulness of the technique comes about because the integral in Eqn. 11-7 can be evaluated in analytic form. Results for the two components of $\vec{E}$ are

$$E_x = \frac{1}{4\pi\varepsilon_0} \sum_{i=1}^{N} \lambda(x_i + .5a/n) \left[\frac{1}{\left[(x-x_{i+1})^2+y^2\right]^{\frac{1}{2}}} - \frac{1}{\left[(x-x_i)^2+y^2\right]^{\frac{1}{2}}} \right] \qquad (11-8)$$

and

$$E_y = \frac{1}{4\pi\varepsilon_0} \frac{1}{y} \sum_{i=1}^{N} \lambda(x_i + .5a/N) \left[\frac{x_{i+1}-x}{\left[(x-x_{i+1})^2+y^2\right]^{\frac{1}{2}}} - \frac{x_i-x}{\left[(x-x_i)^2+y^2\right]^{\frac{1}{2}}} \right] . \qquad (11-9)$$

The flow chart is shown in Fig. 11-3. The following storage allocations are made:

- X1 length of the line,
- X2 number of segments,
- X3 x coordinate of field point,
- X4 y coordinate of field point,
- X5 x component of field,
- X6 y component of field,
- X7 charge density,
- X8 x coordinate of the charge point,

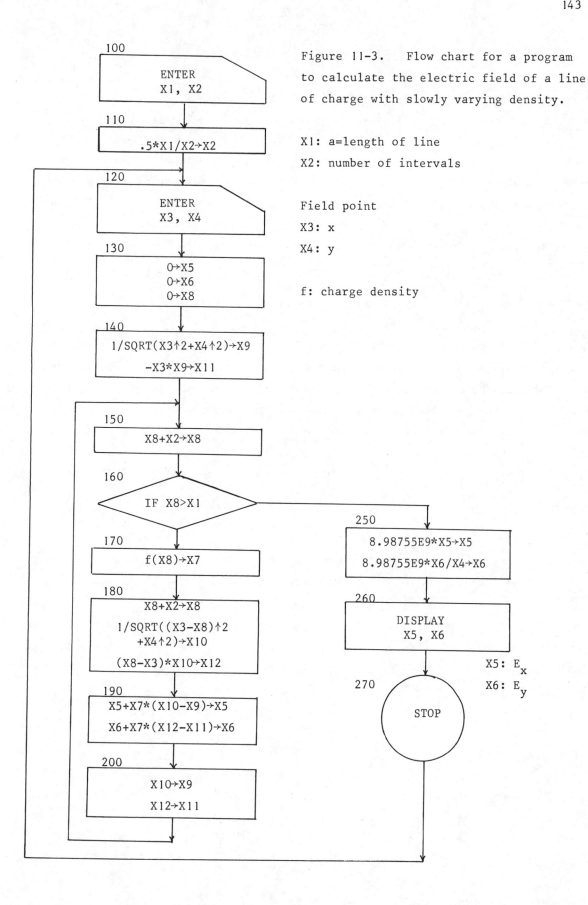

Figure 11-3. Flow chart for a program to calculate the electric field of a line of charge with slowly varying density.

X1: a=length of line
X2: number of intervals

Field point
X3: x
X4: y

f: charge density

X5: E_x
X6: E_y

$$X9 \quad 1/[(x-x_i)^2+y^2]^{\frac{1}{2}} \text{ for segment i,}$$
$$X10 \quad 1/[(x-x_{i+1})^2+y^2]^{\frac{1}{2}} \text{ for segment i,}$$
$$X11 \quad (x_i-x)/[(x-x_i)^2+y^2]^{\frac{1}{2}} \text{ for segment i,}$$
and
$$X12 \quad (x_{i+1}-x)/[(x-x_{i+1})^2+y^2]^{\frac{1}{2}} \text{ for segment i.}$$

When the program enters the loop at line 150, X8 contains the value of x for the beginning of the segment. That is, it contains x_i for segment i. It is immediately incremented by .5a/N, so it contains x for the center of the segment. The charge density is evaluated for this point, at line 170. X8 is again incremented by .5a/N, so it contains x_{i+1}, the coordinate for the end of the segment. At line 180, the quantity $1/[(x-x_{i+1})^2+y^2]^{\frac{1}{2}}$ is calculated and stored in X10 and the quantity $(x_{i+1}-x)/[(x-x_{i+1})^2+y^2]^{\frac{1}{2}}$ is calculated and stored in X12.

At this point in the calculation, X9 contains $1/[(x-x_i)^2+y^2]^{\frac{1}{2}}$ and X11 contains $(x_i-x)/[(x-x_i)^2+y^2]^{\frac{1}{2}}$, both calculated previously. For the first segment, these quantities are calculated at line 140, prior to entering the loop. For this segment $x_i=0$. For other segments, we make use of the fact that these quantities, evaluated for the beginning of the segment, are precisely the same as the quantities calculated for the end of the previous segment and stored in X10 and X12, respectively. All that is required is to transfer the content of X10 to X9 and the content of X12 to X11. This is done at line 200.

At line 190, the contribution of the segment to the electric field is calculated and added to previously calculated contributions. For the x component, the factor $1/4\pi\epsilon_0$ is omitted while, for the y component, the factor $1/4\pi\epsilon_0 y$ is omitted.

If, at line 160, X8>X1, then all segments have been considered and the machine goes to line 250 where the appropriate factors ($1/4\pi\epsilon_0$ for E_x and $1/4\pi\epsilon_0 y$ for E_y) are applied. The field is then displayed and, after restarting, the machine returns to line 120 to consider a new field point.

11.3 Exercises

1. Two charges are located on the x axis: $q_1=6\times 10^{-9}$ C is at $x_1=-0.03$ m and $q_2=3\times 10^{-9}$ C is at $x_2=0.03$ m.

a. Calculate the electric field at the following points along the y axis: y=0, 0.05 m, 0.1 m, 0.15 m, and 0.2 m.

b. Draw a diagram showing the positions of the charges and, at each field point of part a, draw an arrow in the direction of the electric field at that point. The length of the arrow at any point should be proportional to the magnitude of the electric field at that point.

2. Consider the two charges of exercise 1.
   a. Find the electric field at the following points on the y axis: y=-0.05 m, -0.1 m, -0.15 m, and -0.2 m. To the diagram of exercise 1, add arrows which represent the field at these points.
   b. What is the relationship between the x component of the field at y=+0.05 m and the x component of the field at y=-0.05 m? What is the relationship between the y components at these points? Do the same relationships hold for the field at other pairs of points?

3. Two charges are located on the x axis: $q_1 = -3 \times 10^{-9}$ C is at $x_1 = -0.075$ m and $q_2 = 3 \times 10^{-9}$ C is at $x_2 = 0.075$ m.
   a. Calculate the electric field at the following points on the line y=0.03 m: x=-0.15 m, -0.1 m, -0.05 m, 0, 0.05m, 0.1 m, and 0.15 m.
   b. On a piece of graph paper, show the positions of the charges and, at each of the field points of part a, draw an arrow which depicts the direction and magnitude of the electric field at that point.
   c. By considering the fields of the individual charges, explain qualitatively why the y component of the field is negative for field points with negative x coordinates, zero for x=0, and positive for field points with positive x coordinates.
   d. The x component of the electric field reverses sign twice in the region considered. Why? Explain in terms of the fields of the individual charges.
   e. Without making a new calculation, draw field vectors at as many points as you can along the line y=-0.03 m.

4. Two charges are located on the x axis: $q_1 = 3 \times 10^{-9}$ C at $x_1 = -0.075$ m and $q_2 = 3 \times 10^{-9}$ C at $x_2 = 0.075$ m.

a. Calculate the electric field at the following points on the line y=0.03 m: x=-0.15 m, -0.1 m, -0.05 m, 0, 0.05 m, 0.1 m, and 0.15 m.

b. On a piece of graph paper, show the positions of the charges and, at each of the field points of part a, draw an arrow which depicts the direction and magnitude of the electric field at that point.

c. Find a charge and its location so that when its field is added to the field of the two charges, the total field at x=0.15 m, y=0.03 m vanishes. There are many answers to this question.

5. Four particles, each with charge $q=4.5\times 10^{-9}$ C are placed at the corners of a square with side d=0.08 m.

   a. Find the electric field at the 9 points which, along with two corners, divide the diagonal into 10 equal segments.

   b. Draw a diagram which shows the positions of the charges and the electric field at the points of part a. Place arrows representing the field along side the diagonal so that they do not interfere with each other.

6. A line of charge runs along the x axis from the origin to x=0.1 m. Suppose the line contains 5.5×10^{-9} C of charge, distributed uniformly (λ=constant).

   a. Use the program of Fig. 8-1, suitably modified, and 2 segments, to calculate the y component of the electric field at x=0.05 m, y=0.05 m.

   b. Repeat the calculation with N=4.

   c. Continue to double the number of segments and calculate the field until two successive calculations yield results which agree to 3 significant figures.

   d. Repeat the calculation using the program of Fig. 11-3 and 1 segment.

7. Consider the line of charge of exercise 6.

   a. Use the program of Fig. 8-1, suitably modified, and 2 segments, to calculate the y component of the electric field at x=0.05 m, y=10 m.

   b. Double the number of segments and continue to double it until two successive calculations yield results which agree to 3 significant figures.

   c. Why are fewer segments needed for this calculation than for the calculation of exercise 6?

8. Consider the line of charge of exercise 6.
   a. Use the program of Fig. 8-1, suitably modified, and 2 segments to calculate the y component of the electric field at x=0.05 m, y=0.005 m.
   b. Double the number of segments used and continue to double it until two successive calculations yield results which agree to 3 significant figures.
   c. Why are more segments needed for this calculation than for the calculation of exercise 6?

9. Consider the line of charge of exercise 6.
   a. Calculate the electric field at points along the line y=0.05 m. Take points every 0.02 m from x=-0.06 m to x=0.1 m. Use the program of Fig. 11-3.
   b. Draw a diagram which shows the electric field vector at each of the field points of part a. For each point, draw an arrow in the direction of the electric field, with length proportional to the magnitude of the electric field.
   c. Explain qualitatively, in terms of the fields of individual segments of the line, why the sign of the x component of the field is different for field points on different sides of the line's midpoint.

10. Consider a line of charge running from x=0 to x=0.1 m along the axis. Charge is distributed with density given by $\lambda = Ax^2$, where A is a constant. The line contains as much charge as the line of exercise 6 (5.5×10^{-9} C).
    a. Show that
    $$\lambda = 1.65 \times 10^{-5} x^2 \text{ C/m}$$
    for x measured in meters.
    b. Find the electric field at the same field points as were used in exercise 9. Use the program of Fig. 11-3 and a sufficient number of segments to obtain 3 significant figure accuracy.
    c. Draw a diagram which shows the electric field vector at each of the field points of part b.
    d. Compare the fields found in exercise 9 with those of part b of this exercise. For points to the left of the distribution, both the x and the y components of the field are larger in magnitude for the uniform distribution than for the non-uniform distribution. For 0<x<0.05 m, the x component of the field

is larger for the non-uniform distribution than for the uniform distribution. On the right side of the line of charge, the opposite is true. Explain these results in terms of the contributions of individual segments.

11. a. Use $\lambda = 1.65 \times 10^{-5} \, x^2$ C/m and calculate the electric field at points along the line $x=0.05$ m. Take $y=0.1$ m, 1 m, 10 m, 100 m, and 1000 m.
   b. Use $\lambda = 5.5 \times 10^{-8}$ C/m and calculate the electric field for the same set of field points.
   c. Calculate the electric field of a single charge $q = 5.5 \times 10^{-9}$ C, located at $x=0.05$ m, $y=0$. Take the same field points.

   The net charge is the same for all three cases. For the line distributions, you should notice that the further the observation point is from the distribution, the closer the field resembles that of a point charge.

Chapter 12

ELECTRIC POTENTIAL, FIELD LINES, AND GAUSS'S LAW

Programs are given to calculate the electric potential of a set of discrete charges and a continuous distribution of charge along a line. The root finding programs of Chapter 2 are used to plot equipotential surfaces and the polynomial fit of Chapter 4 is used to show the relationship between the electric field and the potential. In the second section, a program which finds the locations of points along electric field lines is presented and used to plot the field lines of various charge distributions. Finally, Gauss's law is demonstrated both by counting field lines, previously plotted, and by actually carrying out detailed calculations. This chapter is intended to be used in conjunction with Chapters 27, 28, and 29 of PHYSICS, Chapters 24, 25, and 26 of FUNDAMENTALS OF PHYSICS, or similar material in another text.

12.1 Electric Potential

The force of the electrostatic field on a charged particle is a conservative force, so it is possible to define a potential energy function for the interaction. It is a function of the position of the particle, defined so that when the particle moves from $\vec{r}_1$ to $\vec{r}_2$ along any path, the electric field does work which is given by the difference $U(\vec{r}_1)=U(\vec{r}_2)$ in the value of the function at the two places.

Instead of potential energy, it is usually convenient to deal with the electric potential: the potential energy of a charge q divided by q. The result is independent of the charge q and, once one knows the potential, one can find the potential energy of a charge by computing the product of charge and potential.

Only differences in potential have physical meaning. It is possible to choose the potential to be zero at some arbitrary point.

In general, the potential at $\vec{r}$ is defined by the line integral

$$V(\vec{r}) = -\int_C \vec{E} \cdot d\vec{r} \qquad (12-1)$$

where the path C extends from the point $\vec{r}_0$, where the potential is zero, to $\vec{r}$. Since the force is conservative, every path with the same end points gives the same value of V and the choice of path is only a matter of computational convenience.

For a single charge q located at $\vec{r}\,'$, the integral can be evaluated, with the result

$$V(\vec{r}) = \frac{1}{4\pi\epsilon_0} \frac{q}{|\vec{r}-\vec{r}\,'|} \qquad (12-2)$$

where the potential has been chosen to be zero for large r. That is, the zero of potential is at infinity.

Note that V is a scalar. The denominator contains the distance from the charge to the field point.

For a collection of charges, with q_i at $\vec{r}_i$, the total potential is the sum of the individual potentials due to the charges taken one at a time:

$$V(\vec{r}) = \frac{1}{4\pi\epsilon_0} \sum_i \frac{q_i}{|\vec{r}-\vec{r}_i|} . \qquad (12-3)$$

For continuous distributions of charge, the sum becomes an integral. If the charge is distributed along the x axis from x=0 to x=a with linear density λ,

$$V(\vec{r}) = \frac{1}{4\pi\epsilon_0} \int_0^a \frac{\lambda(x')\,dx'}{|\vec{r}-\vec{r}\,'|} = \frac{1}{4\pi\epsilon_0} \int_0^a \frac{\lambda(x')\,dx'}{\left[(x-x')^2+y^2\right]^{\frac{1}{2}}} \qquad (12-4)$$

where we have assumed the field point is in the x,y plane.

If the electric field is given as a function of position, then the potential at any point can be found by using Eqn. 12-1. This integral is the same, except for the factor q, as the integral for the potential energy associated with a force and the strategy outlined in section 8.6 can be followed.

The program of Fig. 8-1 can be used. First choose a path from the reference point (where V=0) to the point $\vec{r}$. If x is chosen to be the variable of integration, the function evaluated at lines 130, 170, and 180 of the integration program is

$$-\left[E_x + (dy/dx)E_y\right] \qquad (12-5)$$

where dy/dx is the slope of the path at the integration point. X1 is then the x coordinate of the beginning of the path, X2 is the x coordinate of the end of the path, and X3 is the number of intervals used in the integration.

On the other hand, if the positions for a series of discrete charges is given, we use Eqn. 12-3 and, if a line distribution of charge is given, we use Eqn. 12-4.

Fig. 12-1 shows the flow chart for a program to calculate the potential due to a collection of discrete charges. The charges are located in the x,y plane and the potential is calculated for a field point in that plane. The charges and their positions are stored as for the program of Fig. 11-2 and the program is meant to follow line 140 of that program. Storage allocation is exactly the same except that the potential is now stored in X5 instead of the x component of the electric field and storage location X6, which previously held the y component of the electric field, is now not used.

Fig. 12-2 shows the flow chart for a program to calculate the potential due to a line of charge which lies between x=0 and x=a on the x axis and has linear density λ. The program is modelled after the program of Fig. 8-1 with $\lambda/|\vec{r}-\vec{r}'|$ as the integrand.

The lower limit of the integral is 0 and is not entered as part of the input data. Instead, 0 is placed in X1 at line 140. X2 contains a and X3 contains the number of intervals to be used in the integration. X10 and X11 contain the x and y coordinates, respectively, of the field point. Since y enters into Eqn. 12-4 only in the form y^2, it is squared immediately on entry and the square is stored in X11. This saves some running time since y^2 need not be recomputed for each interval. The function f which appears at lines 140, 180, and 190 is the charge density λ. All other storage allocations are exactly the same as for the program of Fig. 8-1.

The logic of the program and the sequence of statements are also the same with one exception. After the potential is displayed, at line 210, and the machine restarted, it returns to line 120 to accept the coordinates of another field point.

We now use these programs to investigate some properties of the electric potential.

We can easily check by numerical means to see if Eqn. 12-2 gives the same result as Eqn. 12-1 for a single charge.

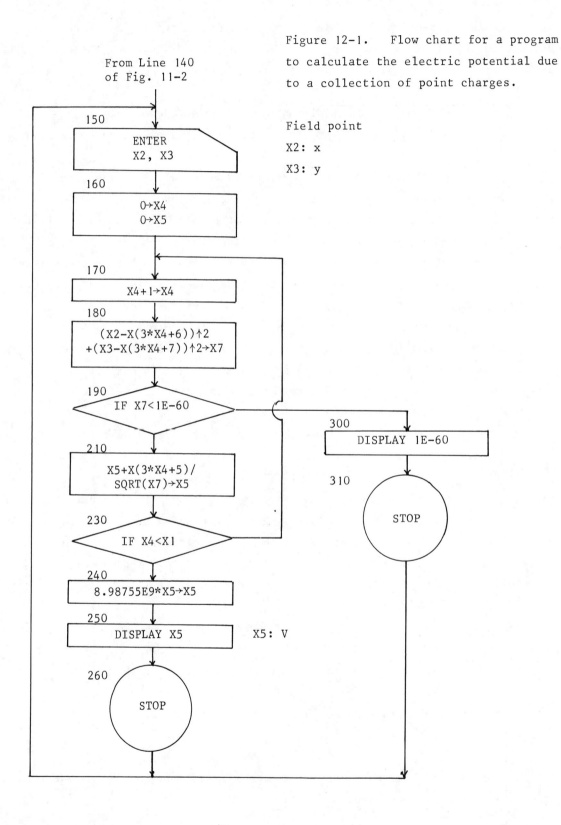

Figure 12-1. Flow chart for a program to calculate the electric potential due to a collection of point charges.

Field point
X2: x
X3: y

X5: V

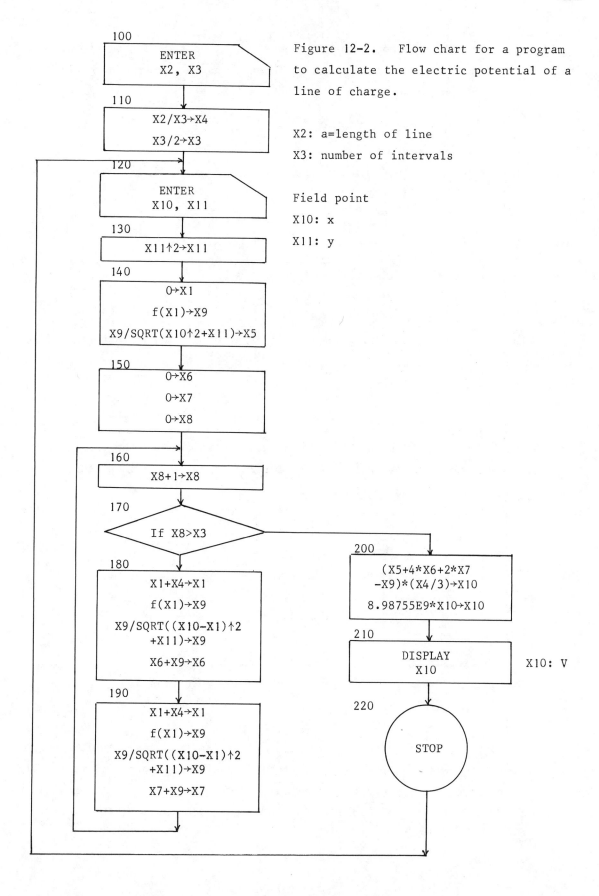

Figure 12-2. Flow chart for a program to calculate the electric potential of a line of charge.

X2: a=length of line
X3: number of intervals

Field point
X10: x
X11: y

X10: V

Problem 1. Consider the point charge $q=3.6\times10^{-9}$ C located at the origin. Use the following two methods to calculate the difference in the electric potential for the two points $x_1=5$ m and $x_2=8.5$ m, both on the x axis.

a. Evaluate

$$V(\vec{r}_2)-V(\vec{r}_1) = (q/4\pi\epsilon_0)\left(\frac{1}{x_2} - \frac{1}{x_1}\right).$$

b. Use the program of Fig. 8-1 to evaluate

$$V(\vec{r}_2)-V(\vec{r}_1) = -\int_5^{8.5} E_x \, dx$$

with $E_x = (q/4\pi\epsilon_0)(1/x^2)$. Use N=8.

For the field of a point charge, the potential difference between two points depends only on the distance of these points from the charge and not in any other way on the locations of the points.

Problem 2. Consider the point charge $q=3.6\times10^{-9}$ C located at the origin. Use the following two methods to calculate the difference in the electric potential for the two points $x_1=5$ m, $y_1=0$ and $x_2=0$, $y_2=8.5$ m.

a. Evaluate

$$V(\vec{r}_2)-V(\vec{r}_1) = (q/4\pi\epsilon_0)\left(\frac{1}{y_2} - \frac{1}{x_1}\right).$$

b. Use the program of Fig. 8-1, as modified in section 8.4, to evaluate

$$V(\vec{r}_2)-V(\vec{r}_1) = -\int \vec{E}\cdot d\vec{r}.$$

Take the path of integration to be the straight line from $\vec{r}_1$ to $\vec{r}_2$, along which $y=8.5(5-x)/5$. Over this path, x decreases and y increases.

For any given charge configuration, there exist surfaces on which the potential is constant. For example, there is a surface for which the potential is 5 volts at every point, another for which the potential is 10 volts, etc. If a charge is carried from a point on one surface to another point on the same surface, the electric field does zero work during the motion. When a charge q goes from one surface to another, however, the electric field does work in the amount $-q\Delta V$, where ΔV is the amount by which the potential of the second surface is higher than the potential of the first.

For the field created by a point charge, equipotential surfaces are spheres centered on the charge, the sphere of radius R being the locus of points with potential $V=(q/4\pi\epsilon_0)(1/R)$ relative to the potential at ∞. The following problem gives a technique for finding the intersections of equipotential surfaces with the x,y plane for more complicated charge distributions.

<u>Problem 3.</u>   Charge $q_1 = -1.2 \times 10^{-9}$ C is at the origin and charge $q_2 = 2.5 \times 10^{-9}$ C is at x=0, y=0.5 m in the x,y plane. On a piece of graph paper, mark axes which run from -5 m to +5 m in both the x and y directions. Mark the positions of the charges.
a. Search along the line x=0 for points at which V=5 volts relative to the potential at ∞. To do this, write

$$V = \frac{1}{4\pi\epsilon_0}\left[\frac{q_1}{\left[x^2+y^2\right]^{\frac{1}{2}}} + \frac{q_2}{\left[x^2+(y-.5)^2\right]^{\frac{1}{2}}}\right],$$

place x=0, and consider the result to be a function of y. Use one of the root finding programs of Chapter 2 to find the root of

$$f(y) = \frac{1}{4\pi\epsilon_0}\left[\frac{q_1}{|y|} + \frac{q_2}{|y-.5|}\right] - 5 \ .$$

There are two such roots. Search from y=-4 m to y=+4 m but avoid y=0 and y=0.5 m where V blows up. On the graph paper, put dots at the positions of the roots. Repeat for other values of x in intervals of 0.25 m until no points are found for some x. For each root, put a dot at the location on the graph. Plot the equipotential surface by connecting the dots. Since $V(-x,y)=V(x,y)$, it is not necessary to carry out separate calculations for negative values of x.

Notice that the equipotential surface cuts between the charges. Notice also that we can find any value of the potential we choose at some point between the charges. The potential is $-\infty$ at the negative charge and $+\infty$ at the positive charge and it varies continuously between. For any potential, the equipotential surface must cut between the charges. In spite of this, the equipotential surfaces at large distances (small V) must be spheres centered on the origin. This is because, at large distances, the charge distribution produces a field which is like that of a single charge of 1.3×10^{-9} C. We can see how these statements are consistent with each other by plotting the 3 volt equipotential surface.

b. Place x=0 in

$$f(y) = \frac{1}{4\pi\epsilon_0} \left[\frac{q_1}{\left[x^2+y^2\right]^{\frac{1}{2}}} + \frac{q_2}{\left[x^2+(y-.5)^2\right]^{\frac{1}{2}}} \right] - 3$$

and find the roots. There are four of them. Repeat for other values of x in intervals of 0.25 m until no points are found for a given x. Plot the equipotential surface. In some regions, narrower intervals might be helpful.

For V=3 volts, there are two surfaces. The larger surrounds both charges while the smaller surrounds only one of the charges.

c. At what value of V does the change from a single to a double surface occur? Answer this question by considering the function

$$f(y) = \frac{1}{4\pi\epsilon_0} \left[\frac{q_1}{|y|} + \frac{q_2}{|y-.5|} \right] - V ,$$

which gives the intersection with the y axis of the equipotential surface for potential V. For some values of V, there are two roots with $-5<y<0$ while, for others, there are no roots. Use trial and error to find the value of V for which the transition occurs. It is instructive to make a table of the roots for various values of V as you search for the transition.

At any point on an equipotential surface, the electric field is perpendicular to the surface. Furthermore, at places where two neighboring equipotential surfaces (5 volts and 6 volts, say) are close together, the magnitude of the electric field is greater than at places where they are far apart.

Problem 4. Consider the charge configuration of problem 3.
a. By direct calculation, using the program of Fig. 11-1 or 11-2, find the electric field at the point farthest from the charges where the 5 volt equipotential

surface crosses the x axis. On the graph you made for problem 3, draw the field vector at this point and verify, to within drawing error, that the field is perpendicular to the surface.

b. The magnitude of $\vec{E}$ is related to the potential by $E=|dV/ds|$ where s is the distance between the equipotential surfaces for V and V+dV. Plot the 6 volt surface in the neighborhood of the point used for part a. On the graph, measure the distance between the 5 volt and 6 volt surfaces and calculate $|\Delta V/\Delta s|$ where $\Delta V=1$ volt. Compare with E, as calculated from the components found in part a. The distance is measured along a line perpendicular to the surface, the line of $\vec{E}$.

The statement in part b of the above problem can be generalized. The x component of the electric field at a point is the negative of the derivative of the potential with respect to x, evaluated at that point. When the derivative is taken, y and z are held constant. In the notation of the calculus,

$$E_x = - \partial V/\partial x \qquad (12-6)$$

where the special derivative symbol, representing what is known as the partial derivative, means that only x (not y and z) is varied in taking the derivative. Similarly,

$$E_y = - \partial V/\partial y \qquad (12-7)$$

and

$$E_z = - \partial V/\partial z. \qquad (12-8)$$

The derivative $\partial V/\partial x$ can be evaluated by finding the potential at 5 points along a line parallel to the x axis, fitting it to a polynomial, and evaluating the derivative of the polynomial.

Problem 5. Consider the following charge configuration: $q_1=-1.63\times 10^{-9}$ C at the origin and $q_2=3.73\times 10^{-9}$ C on the y axis at y=0.45 m.

a. Calculate the potential for the following five points on the line y=2 m in the x,y plane: x=-1.96 m, -1.98 m, -2 m, -2.02 m, and -2.04 m. Fit to a fourth order polynomial with x=-2 m as the central point and evaluate the x component of the electrical field at x=-2 m, y=2 m.

b. Use the same method to calculate the y component of the electric field at the same field point. Use the points x=-2 m with y=1.96 m, 1.98 m, 2 m, 2.02 m, and 2.04 m.

c. Use the program of either Fig. 11-1 or 11-2 to calculate the electric field at x=-2 m, y=2 m. Compare the results with those of parts a and b.

<u>Problem 6.</u> A line of charge runs along the x axis from x=0 to x=0.1 m. Charge is distributed with density given by

$$\lambda(x) = 1.83 \times 10^{-5} \sqrt{x} \quad C/m$$

with x in meters.

a. Use the program of Fig. 12-2, with N=64, to find the electric potential at the following points on the line y=0.2 m, z=0: x=-0.06 m, -0.03 m, 0, 0.03 m, and 0.06 m.
b. Use these results to find the derivative $-\partial V/\partial x$ at x=0.
c. Use the program of Fig.12-2 to find the electric potential at the following points on the y axis: y=0.192 m, 0.196 m, 0.204 m, and 0.208 m. Use N=150.
d. Use the results to find the derivative $-\partial V/\partial y$ at y=0.2 m. You will also need the potential at y=0.2 m. This was found in part a.
e. Use the program of Fig. 11-3, with N=64, to find the electric field at x=0, y=0.2 m and compare with the results of parts b and d.

Agreement can be improved significantly if more intervals are used to answer parts a and c.

12.2 Field Lines

Field lines are drawn to show the direction and, in a qualitative way, the magnitude of the electric field. At any point, the electric field is tangent to the field line through that point and lines are closer together in regions of high field, further apart in regions of low field. Lines originate and end at charges, being directed away from a positive charge and toward a negative charge. It is easy to see why they are useful in visualizing the electric field. We shall learn how to plot field lines for certain charge configurations.

In order to plot electric field lines, it is usual to start at a point close

to one of the charges. There the field line is along the line which joins the point and
the charge. The electric field at the point is calculated and the result is used to
locate a neighboring point on the same field line. The neighboring point is a short
distance away in the direction of the electric field vector. This is an approximation,
of course, which is less and less in error the shorter the distance.

We shall deal with charges, fields, and field lines in the x,y plane. Suppose
the electric field at the point with coordinates x and y has components E_x and E_y and
we wish to find another point on the line which goes through x,y. $(E_x/E)\hat{i} + (E_y/E)\hat{j}$
is a unit vector tangent to the line and

$$x + (E_x/E)\Delta s, \qquad y + (E_y/E)\Delta s \qquad (12-9)$$

are coordinates of a point Δs distance from x,y on the line tangent to the field line.
If Δs is small enough, the sequence of points generated by the repeated use of Eqn. 12-9
all lie on the same field line.

A flow chart for a program to generate a sequence of points on a field line is
given in Fig. 12-3. The electric field is calculated in exactly the same manner as it
is in the program of Fig. 11-2. X1 contains the number of charges to be considered
and X4 counts the charges. The charge is stored in X(3*X4+8), its x coordinate in
X(3*X4+9), and its y coordinate in X(3*X4+10). Once the charge information is entered
and stored, other data is entered at line 150. X2 and X3 contain the x and y
coordinates, respectively, of a point on the field line. Their initial values are
entered and thereafter new values are computed as the field line is traced out. X8
contains Δs, which is, to some extent, arbitrary. Too small and time is lost; too
large and accuracy is lost. It is not worthwhile to display, copy, and plot every point
calculated. X9 contains an integer which tells the machine how many points should be
calculated and not displayed before the next point is displayed. Only the last one of
each X9 points is displayed.

The x component of the field is stored in X5, the y component in X6, and the
magnitude in X7. The factor $1/4\pi\epsilon_0$ is omitted since it does not change the ratios in
Eqn. 12-9. At line 230, the coordinates of the new point are calculated. These are
then used as the starting values for the calculation of the next point.

When the desired field line has been plotted, the machine should be returned to

Figure 12-3. Flow chart for a program to calculate the coordinates of points along electric field lines.

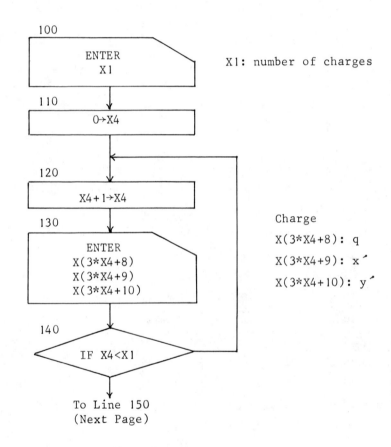

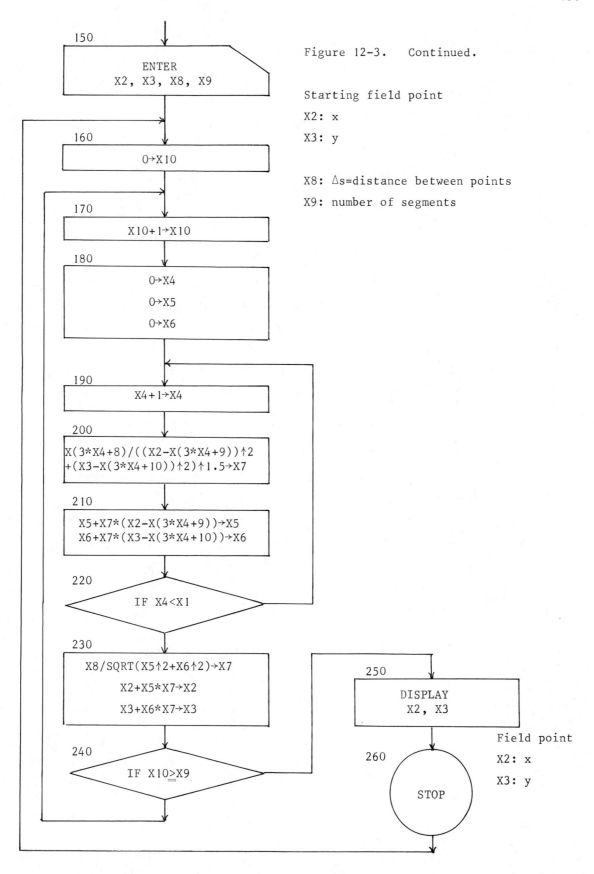

Figure 12-3. Continued.

Starting field point
X2: x
X3: y

X8: Δs=distance between points
X9: number of segments

Field point
X2: x
X3: y

line 150 to consider another field line.

We now use the program to trace electric field lines for some interesting charge configurations.

Problem 1. Charge $q_1=7.1 \times 10^{-9}$ C is located at the origin and charge $q_2=-7.1 \times 10^{-9}$ C is located on the y axis at y=-0.4 m. Plot 2 field lines. Start one of them at $x=1 \times 10^{-3}$ m, $y=1 \times 10^{-3}$ m and the second at $x=1 \times 10^{-3}$ m, $y=-1 \times 10^{-3}$ m. Take $\Delta s=0.005$ m and plot every 20 points. Connect the plotted points with solid lines.

Problem 2. Four identical charges, each with $q=5.6 \times 10^{-9}$ C, are placed with one at each corner of a square with side a=0.36 m, centered at the origin, with sides parallel to either the x or y axis. For the following, take $\Delta s=0.005$ m and plot every 20 points.
a. Draw 6 field lines, each starting on a circle of radius 0.1 m, centered on the charge at x=0.18 m, y=0.18 m. One line starts parallel to the x axis and the others start at intervals of $\pi/3$ radians around the circle. When a line is within 0.1 m of a charge or more than 0.7 m away from all charges, do not continue it.

One of the lines goes toward the center of the charge distribution where the electric field vanishes. The program does not properly evaluate Eqn. 12-9 in the limit of vanishing field and significance is lost. The line is also lost. Simply follow it toward the center, then stop plotting. It actually turns to run close to the y axis, in the positive y direction.
b. Draw 6 field lines emanating from each of the other charges. This can be done using the data generated for the first charge and a symmetry argument. It is not necessary to calculate new points.

Problem 3. Change the sign of the charges at x=-0.18 m, y=0.18 m and x=0.18 m, y=-0.18 m in the configuration of problem 2. Take $\Delta s=0.005$ m and plot every 20 points.
a. Draw 6 field lines emanating from the same charge used in part a of problem 2. Start them at the same points as the lines for that problem. When a line is within 0.1 m of a charge or more than 0.7 m away from all charges, do not continue it.

b. Using symmetry and the results of part a, draw field lines emanating from the other charges.

The field line diagrams are an excellent means to visualize the electric field. In regions of high field, the lines are close together, as they are near the edges of the square for the configuration of problem 3 or near any of the charges. In regions of low field, the lines are far apart, as they are near the center of the square for the configuration of problem 2.

The lines give a graphic display of the mutual repulsion of the charges of problem 2 and the attraction of adjacent charges in problem 3. Since lines run from positive to negative charges, unlike charges tend to have lines running between them and the field lines look as if they tie the charges together. On the other hand, no lines join like charges, lines emanating from one kind of charge avoid charges of the same kind, and the field lines look as if they were pushing against each other.

Far away from all charges, the distribution has the same electric field as that of a single charge equal to the net charge in the configuration and located at the origin. The field lines are then radially outward, if the net charge does not vanish.

For the charge configuration of problem 2, notice that the lines tend to become more nearly in the radial direction as r increases. Because of our choice of starting points for the lines, they also tend to be arranged with equal angles between adjacent lines. At large r, we expect to see 24 lines, all radially outward from the origin, and with adjacent lines separated by $\pi/12$ radians.

For the configuration of problem 3, the net charge is zero and all lines start and stop at charges within the configuration.

12.3 Gauss's Law

Field lines are also useful for the understanding of Gauss's law. The law itself states that the integral of the normal component of the electric field over any closed surface is numerically equal to the net charge enclosed by the surface, divided by ϵ_0. Agreement in sign occurs if the normal component is assigned a positive sign for electric fields which point outward, from inside the surface toward the outside, and a negative sign for electric fields which point in the opposite direction. Symbolically, the law is

$$\oint \vec{E} \cdot d\vec{S} = q/\varepsilon_0. \tag{12-10}$$

Here q represents the net charge enclosed by the surface and $d\vec{S}$ represents an infinitesimal element of surface area on the surface being considered. It is a vector normal to the surface and is directed outward from the surface at the integration point. The combination $\vec{E} \cdot d\vec{S}$ is simply $E_n \, dS$ where E_n is the component of the electric field in the direction of the outward normal to the surface. The circle on the integral symbol reminds us that the surface is closed: it forms the boundary of a finite volume, which it completely encloses.

It is possible to interpret $\vec{E} \cdot d\vec{S}$ as a number proportional to the net number of lines passing through the area $d\vec{S}$. A line makes a positive contribution to the net number if it passes through in the same direction as $d\vec{S}$ and it makes a negative contribution if it passes through in the opposite direction.

Because there are a finite number of lines, there may be some trouble with this interpretation as $d\vec{S}$ becomes infinitesimally small. We assume there are sufficient lines that there is always a large number passing through $d\vec{S}$. Then $\oint \vec{E} \cdot d\vec{S}$ is a number which is proportional to the net number of lines through the entire surface.

Gauss's law is then the statement that the net number of lines passing through the surface is proportional to the net charge inside. It is as if a fixed number of lines, like strings, are attached to each charge, the number of lines being proportional to the amount of charge. Lines which originate on positive charge inside either run to negative charge also inside or else penetrate the surface in an outward direction. In the first case, they contribute nothing to the net number of lines passing through the surface while, in the second case, they contribute a positive number. Lines which end on negative charge inside either come from positive charge also inside or else they pass through the surface in an inward direction. In the first case, they contribute nothing to the net number of lines through the surface while, in the second case, they contribute a negative number. In any event, the net number of lines through the surface and the integral $\oint \vec{E} \cdot d\vec{S}$ are both proportional to the total charge inside. This statement is true no matter how the charge is distributed inside the surface.

The results for problems 2 and 3 of the last section can be used to demonstrate Gauss's law for two dimensions. In each case, 6 lines emanate from each 5.6×10^{-9} C of charge. In two dimensions, we consider, for example, a square, count the net number of lines cutting the boundary, and correlate that number with the net charge inside. You will have to extend some of the lines you drew.

Problem 1. Consider the charge configuration of problem 2 of the last section.
a. On the field line diagram you made, draw a square with edge 0.5 m (to scale), centered on the x axis at x=0.6 m. All edges are parallel to either the x or y axis. Count the net number of lines passing through the boundary of the square. A line contributes +1 to the sum if it exits the square and -1 if it enters the square. For this square, the net charge inside is zero and the net number of lines penetrating the surface should also be zero.
b. Now draw an identical square, placed so that it surrounds one charge, and again count the net number of lines through the boundary. Note that some of the lines originate on enclosed charge while other lines both enter and leave the square and so contribute zero to the net number penetrating the boundary. Since there is 5.6×10^{-9} C of charge enclosed, there should be a net number of 6 lines leaving the boundary.
c. Now draw an identical square, placed so that it encloses 2 charges. The net number of lines penetrating the surface should now be 12. Note that it does not matter how the square is positioned as long as it encloses 2 charges. Count the net number of lines through the surface.

Problem 2. Consider the charge configuration of problem 3 of the last section.
a. On the field diagram, draw a square with edge 0.5 m (to scale), centered on the x axis at x=0.6 m. The sides are parallel to either the x or the y axis. Count the net number of lines through the boundary. There is no net charge enclosed by the square, so this number should be zero.
b. Now draw an identical square, placed so that it encloses one of the negative charges. Verify that 6 lines pass through the boundary and that they are inward going.
c. Now draw an identical square, placed so that it encloses one positive and one negative charge. There should now be a net of zero lines passing through the boundary since there is zero net charge within the square. Verify this.

Gauss's law can be verified directly by evaluating the integral $\oint \vec{E} \cdot d\vec{S}$ over a closed surface. We consider the special case of a cube bounded by the 6 planes $x=0$, $x=a$, $y=0$, $y=a$, $z=0$, and $z=a$ respectively, as shown in Fig. 12-4. We place a single charge at $\vec{r}' = x'\hat{i} + y'\hat{j} + z'\hat{k}$ and carry out the evaluation of $\oint \vec{E} \cdot d\vec{S}$, one face of the cube at a time. The flow chart shown in Fig. 12-5 is for a program which evaluates the integral for the face $x=a$.

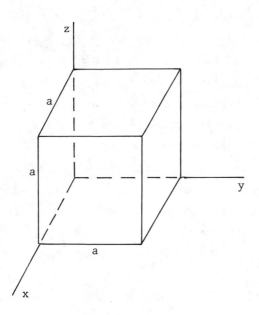

Fig. 12-4. A cube used as the surface of integration for verification of Gauss's law.

For this face, $d\vec{S} = \hat{i}\, dy\, dz$ and

$$\vec{E} \cdot d\vec{S} = (q/4\pi\varepsilon_0)(a-x')\, dy\, dz\, /\, |\vec{r}-\vec{r}'|^3. \tag{12-11}$$

Here
$$|\vec{r}-\vec{r}'| = \left[(a-x')^2 + (y-y')^2 + (z-z')^2\right]^{\frac{1}{2}} \tag{12-12}$$

and y and z are the coordinates of a point on the face. They are the variables of integration.

To evaluate the integral, we divide the y and z axes from 0 to a into N segments each, evaluate $1/|\vec{r}-\vec{r}'|^3$ at the center of each of the N^2 squares formed, sum the contributions, and multiply the result by $(q/4\pi\varepsilon_0)(a-x')(a/N)^2$. Here we have used the fact that the area of each small square is $(a/N)^2$. The centers of the small squares have coordinates which are given by

$$y = (a/N)(i-.5) \qquad (12\text{-}13)$$

and
$$z = (a/N)(j-.5) \qquad (12\text{-}14)$$

where i and j are integers, each of which successively takes on the values 1, 2, 3, ... N as the N^2 squares are considered.

Storage allocation for the program of Fig. 12-5 is

    X1   charge,
    X2   x coordinate of the charge,
    X3   y coordinate of the charge,
    X4   z coordinate of the charge,
    X5   cube edge,
    X6   number of segments (at line 110), a/N (at line 120 and thereafter),
    X7   y coordinate of integration point on cube face,
    X8   z coordinate of integration point on cube face,

and   X9   used to collect the sum of $1/|\vec{r}-\vec{r}\,'|^3$.

The program has two loops, one within the other. For the outer loop, X7 (containing y) is initialized at $-.5(a/N)$ and then incremented by a/N each time around, to produce the sequence of values specified by Eqn. 12-13. For each value of X7, X8 (containing z) is initialized at $-.5(a/N)$ and then incremented by a/N each time around the inner loop. Thus the values specified by Eqn. 12-14 are produced.

At line 180, $1/|\vec{r}-\vec{r}\,'|^3$ is calculated and added to previous contributions. When all area elements have been considered, the machine goes to line 200 where the sum is multiplied by $(q/4\pi\epsilon_0)(a-x\,')(a/N)^2$ to produce the value of the integral.

The program can easily be modified to calculate the contributions of the other 5 faces of the cube. For the x=0 face, the outward normal is in the negative x direction, $d\vec{S}=-\hat{i}\,dy\,dz$, and

$$\vec{E}\cdot d\vec{S} = -(q/4\pi\epsilon_0)(-x\,')\,dy\,dz\,/\,|\vec{r}-\vec{r}\,'|^3. \qquad (12\text{-}15)$$

Here $|\vec{r}-\vec{r}\,'| = \left[(x\,')^2+(y-y\,')^2+(z-z\,')^2\right]^{\frac{1}{2}}$ since x=0 on this face. The statement at

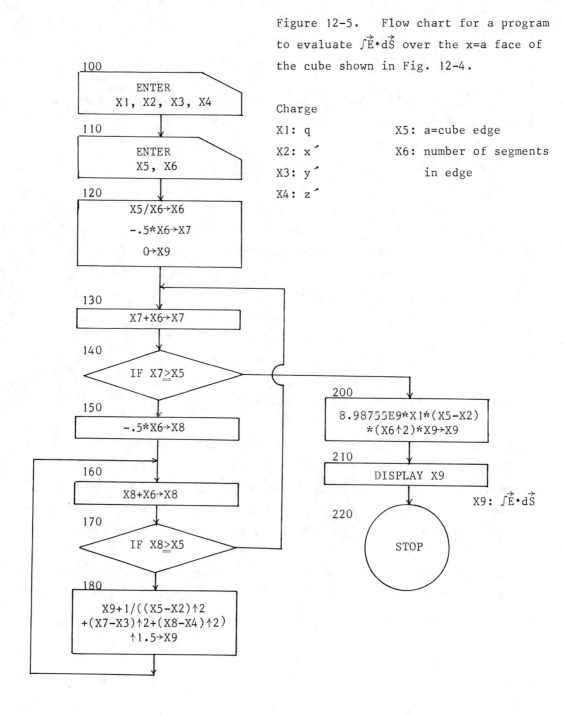

Figure 12-5. Flow chart for a program to evaluate $\int \vec{E} \cdot d\vec{S}$ over the x=a face of the cube shown in Fig. 12-4.

Charge
X1: q   X5: a=cube edge
X2: x´   X6: number of segments
X3: y´   in edge
X4: z´

X9: $\int \vec{E} \cdot d\vec{S}$

line 180 should be changed to read

$$X9+1/(X2\uparrow 2+(X7-X3)\uparrow 2+(X8-X4)\uparrow 2)\uparrow 1.5 \to X9$$

and the statement at line 200 should be changed to read

$$8.98755E9*X1*X2*(X6\uparrow 2)*X9 \to X9$$

To evaluate the contributions of the y=0 and y=a faces, we assign x to X7 and z to X8. To evaluate the contributions of the z=0 and z=a faces, we assign x to X7 and y to X8. These are the coordinates of the integration point. Storage locations for the coordinates of the charge are not changed. The following table shows how lines 180 and 200 should read to calculate the contribution of each of these faces.

Face	Line 180	Line 200
y=a	$X9+1/((X7-X2)\uparrow 2+(X5-X3)\uparrow 2 +(X8-X4)\uparrow 2)\uparrow 1.5 \to X9$	$8.98755E9*X1*(X5-X3) *(X6\uparrow 2)*X9 \to X9$
y=0	$X9+1/((X7-X2)\uparrow 2+X3\uparrow 2 +(X8-X4)\uparrow 2)\uparrow 1.5 \to X9$	$8.98755E9*X1*X3 *(X6\uparrow 2)*X9 \to X9$
z=a	$X9+1/((X7-X2)\uparrow 2+(X8-X3)\uparrow 2 +(X5-X4)\uparrow 2)\uparrow 1.5 \to X9$	$8.98755E9*X1*(X5-X4) *(X6\uparrow 2)*X9 \to X9$
z=0	$X9+1/((X7-X2)\uparrow 2+(X8-X3)\uparrow 2 +X4\uparrow 2)\uparrow 1.5 \to X9$	$8.98755E9*X1*X4 *(X6\uparrow 2)*X9 \to X9$

In the following problems, these programs are used to verify some statements about Gauss's law.

Computation time for the problems can be shortened considerably by taking advantage of symmetry. The contributions of two or more faces of the cube may be the same because the charge is situated in the same position relative to each face. When this occurs, there is no need to perform the integration for more than one of the faces involved. Also, for some of the faces and for some charge configurations, $\vec{E}$ is perpendicular to $d\vec{S}$ ($\vec{E}$ is parallel to the cube face) and the contribution to the integral is zero. The numerical integration need not be performed if this situation is recognized.

Problem 3. Take a=10 m and evaluate $\oint \vec{E} \cdot d\vec{S}$ for the surface of a cube with charge $q=3.7 \times 10^{-9}$ C inside. Use N=15 for 3 significant figure accuracy.
a. First assume the charge is at the center: x=5 m, y=5 m, z=5 m. Compare the result with q/ε_0.
b. Now place the charge at x=2.5 m, y=3 m, z=6.3 m. Compare the result with q/ε_0.

If there is zero charge within the cube, Gauss's law predicts that $\oint \vec{E} \cdot d\vec{S}=0$, no matter how much charge is outside or how it is distributed on the outside.

Problem 4. Take a=10 m and evaluate $\oint \vec{E} \cdot d\vec{S}$ for the surface of the cube, with charge $q=3.7 \times 10^{-9}$ C outside. Use N=15 for 3 significant figure accuracy.
a. First assume the charge is located at x=15 m, y=5 m, z=5 m.
b. Now place the charge at x=0, y=17 m, z=0.

The results of these calculations might not be exactly zero because of integration error. If N is increased, both results are smaller.

Problem 5. Without carrying out a numerical calculation, but using the statements of the program of Fig. 12-5, give arguments to show why the program produces the following results.
a. If both q_1 and q_2 are charges inside the cube, then $\oint \vec{E} \cdot d\vec{S}$ is the sum of the integrals for each of the charges, evaluated one at a time.
b. If $q_1=-q_2$ and both charges are inside the cube, then $\oint \vec{E} \cdot d\vec{S}=0$, no matter what their positions inside.
c. If q_1 is inside and q_2 is outside the cube, then $\oint \vec{E} \cdot d\vec{S}=q_1/\varepsilon_0$.

Chapter 13
SOLUTION OF SIMULTANEOUS EQUATIONS
WITH APPLICATION TO CIRCUIT PROBLEMS

A method for solving simultaneous linear equations is discussed and a flow chart for its implementation is presented. The method is applied to find currents in DC circuits. As one example, the Wheatstone bridge, both balanced and unbalanced, is investigated. The program given here can also be used to solve for currents in circuits which arise in connection with laboratory experiments. This material supplements Chapters 31 and 32 of PHYSICS, Chapters 28 and 29 of FUNDAMENTALS OF PHYSICS, and similar sections of other texts.

13.1 <u>Simultaneous Equations</u>

For many situations, the application of physical principles leads to several equations containing several unknowns. If the number of unknowns is the same as the number of equations and another condition, to be discussed shortly, is met, the equations can be solved simultaneously for the unknowns.

As an example, two equations which can be solved for the two unknowns x_1 and x_2 are

$$5.3x_1 + 18.1x_2 = 3.5$$
$$1.8x_1 + 3.2x_2 = 1.4 \qquad (13-1)$$

Other examples, which arise in connection with the analysis of electric circuit problems, are given in the next section.

We consider equations which are linear. That is, each term either contains no unknowns or is directly proportional to a single unknown. Terms containing $x_1 x_2$ or x_1^2, for example, cannot occur. We also consider systems of equations for which at least one of the equations contains a non-zero term with no unknowns in it, such as the terms to the right of the equal signs in Eqns. 13-1.

For these systems of equations, a solution exists if the determinant of the coefficients does not vanish. This determinant is constructed so that its first row contains, in order, the coefficients of x_1, x_2, x_3, etc. in the first equation. Its second row contains, in order, the coefficients of x_1, x_2, x_3, etc. in the second equation. Other rows are constructed in a similar manner. For the system of equations given in Eqns. 13-1, the determinant of the coefficients is

$$\begin{vmatrix} 5.3 & 18.1 \\ 1.8 & 3.2 \end{vmatrix} = -15.62$$

Care must be taken if one or more of the unknown quantities do not appear in some of the equations. In that case, the corresponding coefficients are zero.

In order to write the equations in compact form, we number the unknowns and use the symbol x_j to represent unknown number j. We also number the equations and use i as a subscript to represent equation i. Equation i is then written

$$A_{i1}x_1 + A_{i2}x_2 + A_{i3}x_3 + \ldots + A_{iN}x_N = B_i \tag{13-2}$$

where we have assumed N unknowns and used B_i to represent the term which does not contain any unknowns. There must be N equations and i takes on the values 1, 2, ... N. In more compact form

$$\sum_{j=1}^{N} A_{ij}x_j = B_i, \quad i=1,2,3, \ldots N. \tag{13-3}$$

All the coefficients A_{ij} and the numbers B_i are assumed to be given and the goal is to solve these N equations for the N unknown quantities. We use an iteration scheme, known as the Gauss-Seidel method, in which we guess at the solutions, then use the equations to obtain better guesses. This continues until we have a solution to within tolerable error.

We solve the first equation for x_1 in terms of the other unknowns, the second equation for x_2, and, in general, equation i for x_i. The result is

$$x_i = \left[B_i - \sum_{\substack{j=1 \\ j \neq i}}^{N} A_{ij}x_j \right] / A_{ii} \tag{13-4}$$

where the term for which j=i is omitted from the sum.

The first step is to guess solutions x_1, x_2, x_3, ... x_N. These guesses are used to calculate a new x_1, using Eqn. 13-4 with i=1. The new x_1 and the old x_3, x_4, ... x_N are used to calculate a new x_2, using Eqn. 13-4 with i=2. And so it goes until all N unknowns have been recomputed. Then the sequence of steps is carried out again, using the results of the first run. Iteration is continued until two successive runs yield the same results to within acceptable error. Each time around, the most current values of the unknowns are used on the right side of Eqn. 13-4 to calculate the new value of x_i. That value then immediately replaces the old value of x_i.

In order to design a program to carry out this strategy, we need a scheme to store and retrieve the coefficients A_{ij}, the constants B_i, and the solutions x_i. The scheme we use stores in order, the A_{ij} and B_i for i=1, then for i=2 and the succeeding equations. Since we need the first 6 storage locations for the calculation, we start with X7. For example, if there are 3 equations in 3 unknowns, we want the allocation of memory to be

$$A_{11}: X7 \quad A_{12}: X8 \quad A_{13}: X9 \quad B_1: X10$$
$$A_{21}: X11 \quad A_{22}: X12 \quad A_{23}: X13 \quad B_2: X14$$
$$A_{31}: X15 \quad A_{32}: X16 \quad A_{33}: X17 \quad B_3: X18$$

We then place x_1 in X19, x_2 in X20, and x_3 in X21.

A scheme which accomplishes this general arrangement for any number of equations is as follows. A_{ij} is stored in the memory location with number given by (i-1)(N+1)+j+6, B_i is stored in the memory location with number i(N+1)+6, and x_j is stored in the memory location with number N(N+1)+j+6. The reader should check these location numbers for the case of 3 equations in 3 unknowns.

Two convenient short cuts are used in writing the program. As Eqn. 13-4 shows, both A_{ij} and B_i are divided by A_{ii}. It is computationally efficient to perform this division as soon as possible, so, as soon as the A_{ij} and B_i are entered for a given i, each A_{ij} is replaced by A_{ij}/A_{ii} and B_i is replaced by B_i/A_{ii}. A_{ii} is temporarily placed in a different storage location to carry out the division.

Secondly, it is inconvenient and time consuming to omit the j=i term from the sum which occurs in Eqn. 13-4. This requires the use of a conditional transfer statement and it is computationally faster to include a new term $A_{ii}x_i$ with $A_{ii}=0$. The consequence of these considerations is that the original A_{ij} and the original B_i are divided by the original A_{ii}, then A_{ii} is set equal to 0. The sum then includes all terms.

Fig. 13-1 gives the flow chart. The first part, from line 100 through line 220, gives the instructions for entering the coefficients A_{ij} and the constants B_i. The order is A_{i1}, A_{i2}, ... A_{iN}, B_i for a given i. This sequence is followed for each i in turn from i=1 to i=N.

Storage allocation is

    X1 number of equations,

    X2 i, label of the equation being considered,

    X3 j, label of term being considered,

    X4 the combination (i-1)(N+1)+j+6 which gives the number of the storage location for A_{ij} and for B_i when j=N+1,

    X5 the combination (i-1)(N+1)+i+6 which gives the number of the storage location for A_{ii},

and   X6 A_{ii}.

There are three loops in this section of the program. The outer one, from line 120 to line 220, is traversed once for each of the N equations. Each time around, X2, which contains i, is incremented by 1 and the next equation is considered.

The first inner loop, from line 130 to line 160, allows the A_{ij} and B_i for the equation being considered to be entered sequentially. X3 contains j and runs from 1 to N+1 as the loop is traversed N+1 times (the first N times to enter the A_{ij} and the last time to enter B_i). At 140, the storage location for the next number to be entered is calculated and stored in X4.

When all of the A_{ij} and B_i for a given equation have been entered, the machine proceeds to divide each A_{ij} and B_i by A_{ii} and store the result back in the original location. Instructions for this process are in the loop from line 180 to line 200. Again X3 contains j and it runs from 1 to N+1. Prior to entering the loop, at 170, A_{ii} is placed in X6. At line 190, in the loop, A_{ij} is divided by A_{ii} and the old A_{ij}

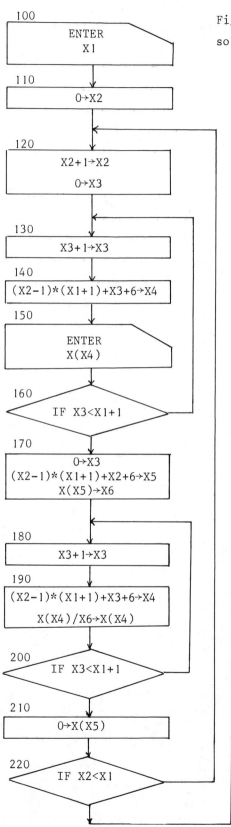
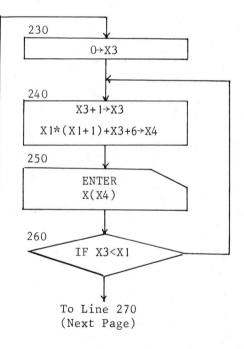

Figure 13-1. Flow chart for a program to solve simultaneous linear equations.

X1: number of equations
X(X4): coefficients in order
$A_{11}, \ldots A_{1N}, B_1$
$A_{21}, \ldots A_{2N}, B_2$
.
.
.
Then quesses to solution in order
$x_1, x_2, \ldots x_N$

To Line 270
(Next Page)

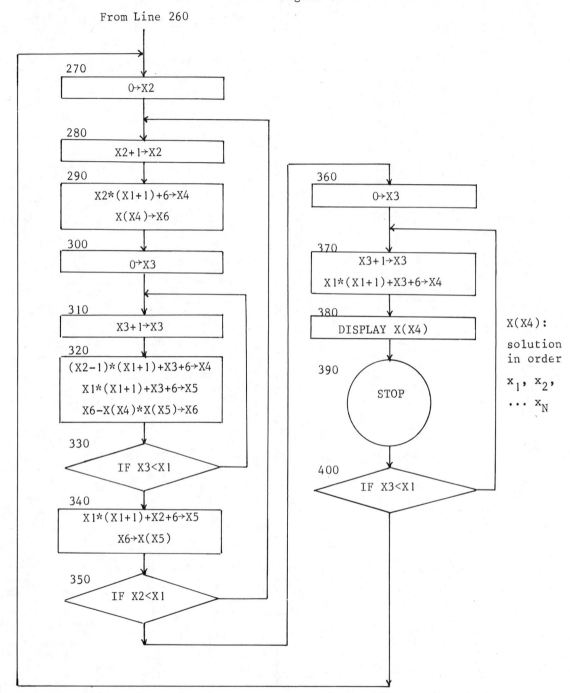

Figure 13-1. Continued.

is replaced by the result. The last time through the loop, B_i is divided by A_{ii} and the old B_i is replaced by the result. At line 210, A_{ii} is set equal to zero, and the machine, at 220, tests to see if all the equations have been considered. If they have, it goes to line 230.

The loop from line 240 to line 260 allows the user to enter initial guesses for the values of the x_j. Here X4 contains the number of the storage location for x_j. If no guesses can reasonably be made, it is usual to set all the x_j equal to zero for the first run.

The sequence of statements from line 270 to line 350 contains two loops. In the outer loop, Eqn. 13-4 is evaluated for each i. X2 contains i and each time around the loop, it is incremented by 1 until all N of the x_i have been computed. The value of the expression on the right side of Eqn. 13-4 is temporarily stored in X6 during the computation process. At line 290, B_i/A_{ii} is entered into X6. The purpose of the loop from line 310 to line 330 is to calculate the individual contributions $A_{ij}x_j/A_{ii}$ to the sum and subtract them from B_i/A_{ii}. X3 contains j and, each time around this inner loop, the machine increments it by 1 and calculates another contribution.

At line 320, X4 contains the number of the storage location for A_{ij}/A_{ii} and X5 contains the number of the storage location for x_j. The product $(A_{ij}/A_{ii})x_j$ is found and the result subtracted from the current content of X6. The result is placed back in X6, ready for subtraction of the next contribution.

When the new value of x_i has been computed, it is placed in the storage location allocated x_i. This occurs at line 340. Here X5 contains the number of the storage location for x_i.

When all N of the x_i have been found, the machine goes to line 360 where a loop is entered to sequentially display the N values. X3 contains j, X4 contains the number of the storage location for x_j, and each x_j is displayed in turn.

Following the display, the machine returns to line 270 to start the next iteration. Note that the previous results are used as the initial guesses. The user should allow the machine to continue to perform iterations until two successive sets of results agree sufficiently closely.

This program must be used with some care. First, the original equations must be ordered so that none of the A_{ii} vanish. If, for example, x_3 does not appear in one of the equations, this equation cannot be used as the third equation of the set.

Sometimes the results do not converge. After many iterations, successive results may differ greatly and may show no sign of getting closer in value. When this occurs, the original set of equations must be modified by adding (repeatedly, perhaps) some of the equations to others or subtracting some of the equations from others. An equation developed in this way replaces one of the original equations. The new system of equations has the same solutions as the original set.

One must have a criterion which will tell when the iteration scheme converges to a solution. We state without proof that the Gauss-Seidel iteration method converges toward the correct solution if, for each equation in the set, the so-called diagonal coefficient (A_{ii}) is larger in magnitude than the sum of the magnitudes of the other coefficients in the equation. That is, for equation i,

$$|A_{ii}| > \sum_{\substack{j=1 \\ j \neq i}}^{N} |A_{ij}| . \tag{13-5}$$

If this inequality is true for all N equations of the set, then the method converges. The condition is independent of the B_i and it does not give any indication of how many iterations are required for convergence. Also, it is a sufficient and not a necessary condition. The method may converge for sets of equations which do not meet the condition.

If the method does not converge, the equations must be rearranged or otherwise manipulated in some fashion to produce a set of equations which have the same solutions as the original set and which meet the convergence criterion.

Sometimes a simple reordering of the equations is sufficient to do the job. For example, consider the equations

$$-2x_1 + 15x_2 + 5x_3 = 17$$
$$8x_1 - x_2 + 3x_3 = -10$$
$$8x_1 + x_2 - 10x_3 = 7$$

Since 2<15+5 and 1<8+3, the inequality given in Eqn. 13-5 is not met by either the first or the second equation of the set. However, an interchange of these two equations produces

$$8x_1 - x_2 + 3x_3 = -10$$
$$-2x_1 + 15x_2 + 5x_3 = 17$$
$$8x_1 + x_2 - 10x_3 = 7$$

and the inequality is satisfied by all three equations.

A general method exists for finding the optimum arrangement. Search the coefficients to find the one which is largest in magnitude and arrange the order of the equations so that this coefficient lies on the diagonal. Now search for the coefficient which is next largest in magnitude. Ignore coefficients in the same equation or multiplying the same unknown as the one found previously. Arrange the equations so that the second largest coefficient is also on the diagonal. Continue this process until there are no more equations to be considered.

For the set of equations above, 15 is the largest coefficient and the equation originally written first is now placed in the second position so that this coefficient is a diagonal coefficient. 10 is the next largest coefficient, so the equation originally written in the third position remains in that position. The second equation must be first in the reordered scheme. That is the only place open for it.

This procedure does not guarantee that the condition given by Eqn. 13-5 is met. Nevertheless, once it has been carried out, the program should be run to test for convergence.

A more difficult case to handle occurs when simple interchange does not bring about convergence. Consider the equations

$$3x_1 + 2x_2 + 2x_3 = 8$$
$$3x_1 + 4x_2 + 3x_3 = 5$$
$$7x_1 + 5x_2 + 3x_3 = 3$$

None of these equations obeys the inequality, but a new set can be found which does. The new set is found by adding multiples of these equations to each other or subtracting multiples from each other.

First we put the equations in optimum order:

$$7x_1 + 5x_2 + 3x_3 = 3$$
$$3x_1 + 4x_2 + 3x_3 = 5$$
$$3x_1 + 2x_2 + 2x_3 = 8$$

Now subtract the second equation from the first and use the result to replace the first. The new set of equations is

$$4x_1 + x_2 = -2$$
$$3x_1 + 4x_2 + 3x_3 = 5$$
$$3x_1 + 2x_2 + 2x_3 = 8$$

Next subtract the third equation from the second and use the result to replace the second:

$$4x_1 + x_2 = -2$$
$$2x_2 + x_3 = -3$$
$$3x_1 + 2x_2 + 2x_3 = 8$$

Both the first and second equations now obey the inequality.

To bring the third equation into line, multiply the first by 3, the third by 4 and subtract the results. Use this to replace the third equation. Then

$$4x_1 + x_2 = -2$$
$$2x_2 + x_3 = -3$$
$$5x_2 + 8x_3 = 38$$

All of these equations satisfy the inequality and we expect the Gauss-Seidel iteration method to converge to a solution for this set of equations.

13.2 Exercises

Use the program of Fig. 13-1 to solve for the unknowns in the following sets of simultaneous equations. If the method does not converge for the equations as they stand, manipulate them to form a set which meets the convergence criterion and which has the same solution as the original set. Then apply the program.

1. $5x_1 + 3x_2 = -6$
    $-2x_1 + 8x_2 = 1$

2. $5x_1 + 3x_2 = -6$
    $8x_1 - 2x_2 = 1$

3. $6x_1 + x_2 - 4x_3 = 7$
    $-3x_1 + 8x_2 - 4x_3 = -2$
    $5x_2 + 3x_2 - 10x_3 = 4$

4. $-15x_1 - 6x_2 + 7x_3 = 9$
    $2x_1 + 3x_2 - 12x_3 = -17$
    $7x_1 + 9x_2 - x_3 = 22$

5. $-15x_1 - 6x_2 + 7x_3 = 9$
    $2x_1 + 3x_2 - 12x_3 = -17$
    $7x_1 + 15x_2 - x_3 + 5x_4 = 22$
    $3x_1 + 7x_4 = -16$

6. $2x_1 - x_2 = 5$
    $2x_2 - x_3 = 7$
    $2x_3 - x_4 = 9$
    $2x_4 - x_1 = 11$

13.3 Direct Current Circuits

Direct current electrical circuits consist of branches, which contain, in general, sites of electromotive force (emf) and resistances:

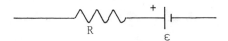

Branches join other branches at junctions and, if current is to flow in any branch, both ends must be joined to other branches to form complete circuits or loops.

In a typical problem, all the emf's and resistances are given and one is asked to find the current in each branch. The first step toward a solution is to arbitrarily assign symbols and directions to the current in each branch. Usually an arrow tells the direction assigned positive current and the symbol i_j is used for the current in branch j. If, upon solution, the current turns out to be negative, that means it flows in the direction opposite to the arrow.

For circuits in which the currents are constant, Kirchoff's laws give a sufficient number of equations to solve for all the currents if the emf's and resistances are given. They are of two forms. The junction equations relate the currents which enter and leave the various junctions. At any junction, the sum of the currents with arrows pointing into the junction equals the sum of the currents with arrows pointing out of the junction. For the junction

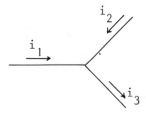

$i_1+i_2=i_3$. This equation is true even if some or all of the currents turn out to be negative.

It is invariably useful to define the symbols for the various currents so that the junction equations are automatically satisfied. For the junction above, for example, the currents can be written

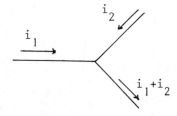

This procedure eliminates one variable (i_3) and one equation from the set of equations needed for the circuit. In the examples and problems that follow, we use this procedure whenever possible.

The second type equation is called a loop equation. One traverses a complete circuit or loop, adds up the potential drops around the loop, and equates the result to zero. When one traverses a source of emf from the negative to the positive terminal, the potential increases by ε. When one traverses a resistor in the direction of the current arrow, the potential decreases by iR where i is the current through the resistor and R is the resistance. Traversals in the opposite direction lead to potential changes of opposite sign in both cases.

For the loop shown below, we start at A and go around in a clockwise direction. The potential at A is taken to be zero and the potential at B is $-i_2 R_1$. At C it is $-i_2 R_1 + \varepsilon$, at D it is $-i_2 R_1 + \varepsilon - (i_2 - i_3) R_2$, and finally back at A, it is $-i_2 R_1 + \varepsilon - (i_2 - i_3) R_2 + (i_1 - i_2) R_3$. This, of course, must vanish and so the loop equation for this loop is

$$-i_2 R_1 + \varepsilon - (i_2 - i_3) R_2 + (i_1 - i_2) R_3 = 0.$$

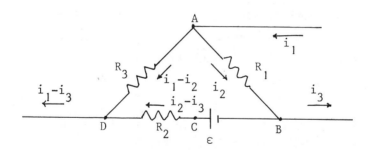

Note that the potential at B is $i_2 R_1$ lower than the potential at A and the potential at C is ε higher than the potential at B. The potential at A is $(i_1 - i_2) R_3$ higher than the potential at D, provided the currents flow as indicated by the arrows. These statements are true in an algebraic sense whether or not the arrows indicate the true flow of current. That is, $V_A - V_B = i_2 R_1$ whether i_2 is positive or negative.

If the current symbols are defined to automatically satisfy the junction equations, then there must be one loop equation for each different current symbol. Care must be taken that every branch is used in at least one loop. Branches can be used in more than one loop, however. If a branch is not used, one of the equations already written must be replaced by a loop equation which makes use of that branch.

The first problem is presented as an uncomplicated example.

Problem 1.   Consider the circuit shown below. Current arrows and labels have been included.

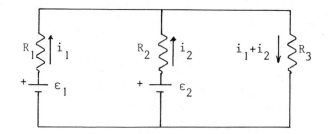

a. Write down two loop equations, one using the loop which contains R_1 and R_3, the other using the loop which contains R_2 and R_3.
b. Take $R_1=R_2=8\ \Omega$, $R_3=12\ \Omega$, $\varepsilon_2=6$ V, and use the program of Fig. 13-1 to solve for i_1 and i_2 for each of the following values of ε_1: 0 V, 3 V, 6 V, and 9 V.
c. For each of the situations of part b, tell which of the batteries is charging and which is discharging. If the actual current flow is through the emf source from the negative to the positive terminal, the battery is discharging. Otherwise it is charging.

The next problem concerns a slightly more complicated circuit.

Problem 2.   Consider the circuit shown. Current arrows and labels have been included.

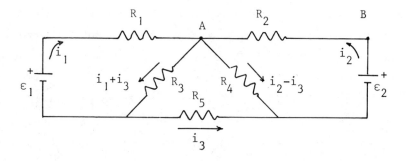

a. Write down three loop equations which can, in principle, be solved for the three unknown currents. Use the loop which contains R_1 and R_3, the loop which contains R_3, R_4, and R_5, and the loop which contains R_2 and R_4.
b. Take $\varepsilon_1 = 10$ V, $R_1 = R_2 = 5\ \Omega$, $R_3 = R_4 = 8\ \Omega$, $R_5 = 12\ \Omega$, and find the three currents for each of the following values of ε_2: 5 V, 7.5 V, 10 V, 12.5 V, and 15 V.
c. For each of the situations of part b, tell if ε_2 is charging or discharging.

Sometimes, one or more of the currents is known. We make the circuit slightly more complicated and consider such a case.

Problem 3. Suppose the points marked A and B on the circuit of problem 2 are also attached to some other circuit as shown below. The external circuit draws current i_4. The complete circuit is not shown.

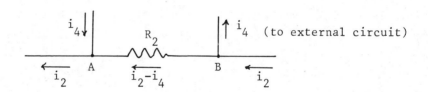

a. Write the loop equations for the same loops as were used in part a of problem 2.
b. Take $\varepsilon_2 = 15$ V, the other quantities as given in problem 2, and solve for i_1, i_2, and i_3 for each of the following values of i_4: 0, 2 A, 4 A, 6 A, 8 A, and 10 A. Obtain 3 figure accuracy.
c. For each of the cases considered, find the potential drop from B to A, $\Delta V = V_B - V_A$.
d. Plot i_4 vs. ΔV.

Notice that i_4 vs. ΔV is a straight line and that ΔV is not zero when i_4 is zero. The entire circuit between A and B, the circuit of problem 2, can be replaced by a source of emf and a resistor in parallel. If the values are chosen correctly, the replacement does not change i_4 vs. ΔV.

Problem 4. Consider the circuit, shown here in part.

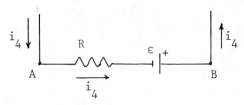

a. Find the values of R and ε so that i_4 vs. ΔV for this circuit matches i_4 vs. ΔV for the circuit of problem 3. Use $V_B - V_A = +\varepsilon - i_4 R$ and choose R and ε by considering the slope and intercept of the graph of problem 3.

This combination of R and ε constitute what is called the Thevenin equivalent circuit corresponding to the original circuit (in problem 2), between A and B. There is a standard technique for finding R and ε. It is as follows.

b. The equivalent emf is the potential drop between the two points when no current is drawn from the circuit. Use the result for i_2 found in problem 2 with $\varepsilon_2 = 15$ V, calculate $i_2 R_2$, and compare the result with the solution to part a.

c. To find R, ε_1 and ε_2 are omitted from the circuit and the equivalent resistance between A and B is found. This is R. When ε_1 and ε_2 are omitted, the circuit is

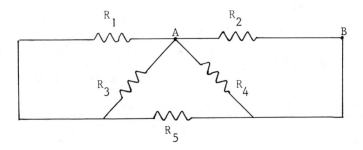

By combining resistances in parallel or series, reduce this network to

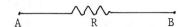

and find the value of R. Compare it with the answer to part a.

Another interesting circuit is the Wheatstone bridge, shown below. It is used to measure resistance and is typical of many different bridge circuits used for various electrical measurements.

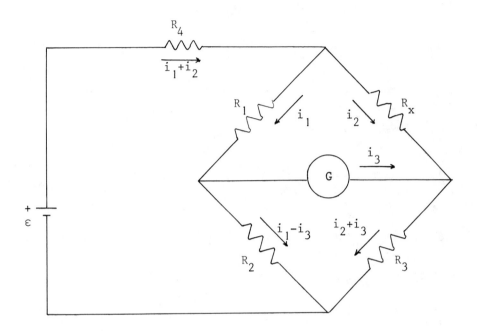

The symbol (G) stands for a galvanometer, an instrument which can detect small currents. The unknown resistor is R_x and the other resistors are variable. They are set so that the galvanometer reads 0, and $i_3=0$. Then

$$R_x = R_1 R_3 / R_2 \qquad (13\text{-}6)$$

and the unknown resistance can be evaluated in terms of known resistances.

We first verify that $i_3=0$ when $R_x = R_1 R_3 / R_2$.

Problem 5.   Take $R_1 = R_2 = 12 \; \Omega$, $R_x = R_3 = 18 \; \Omega$, $R_4 = 3.2 \; \Omega$, and the resistance of the galvanometer $R_5 = 2 \; \Omega$. Use $\varepsilon = 10$ V. Note that the balance condition, Eqn. 13-6, is met.
a. Write three loop equations to be solved for the three unknown currents i_1, i_2, and i_3. Use the loop containing R_1, R_x, and the galvanometer, the loop containing R_2, R_3, and the galvanometer, and the loop containing R_4, R_x, and R_3.

b. Use the program of Fig. 13-1 to solve for i_1, i_2, and i_3 with an accuracy of at least 3 significant figures. If the condition, Eqn. 13-6, is valid the result for i_3 should be zero to within the accuracy of the iteration.

It is usually of some interest to know how sensitive the Wheatstone bridge is. We test the sensitivity by making an error in the setting of one of the resistors, then seeing if the galvanometer can detect the resulting current i_3.

<u>Problem 6.</u> Take $R_1=R_2=12$ Ω, $R_3=19$ Ω, $R_x=18$ Ω, $R_4=3.2$ Ω, $R_5=2$ Ω, and $\varepsilon=10$ V. The balance condition is not met, current i_3 flows through the galvanometer, and Eqn. 13-6 predicts the wrong value for R_x. It predicts 19 Ω, whereas R_x is actually 18 Ω.

Solve for i_1, i_2, and i_3. The galvanometer must be capable of detecting current equal to this value of i_3 if the bridge is to be used to measure R_x with an error less than 1 Ω. Is this an unreasonable current for a galvanometer to detect?

Chapter 14

THE MAGNETIC FIELD

We start by giving programs to calculate the magnetic fields and plot field lines of moving charges. We then consider the fields of currents in circular loops. As practical examples, a pair of Helmholtz coils and a solenoid are investigated as a superposition of fields due to single loops. Finally, Ampere's law is demonstrated numerically. The problems suggested here are meant to illustrate, by way of detailed calculation, the meaning of the law. The material supplements Chapters 33 and 34 of PHYSICS, Chapters 30 and 31 of FUNDAMENTALS OF PHYSICS, or similar material in other texts.

14.1 Magnetic Fields of Moving Charges

A moving charge produces a magnetic field at all points in space. We suppose charge q is located at $\vec{r}\,'$ and has velocity $\vec{v}$. Then the magnetic field $\vec{B}$ it produces at $\vec{r}$ is given by

$$\vec{B}(\vec{r}) = \frac{\mu_0}{4\pi} q \frac{\vec{v} \times (\vec{r} - \vec{r}\,')}{|\vec{r} - \vec{r}\,'|^3} \qquad (14\text{-}1)$$

where $\mu_0/4\pi = 1 \times 10^{-7}$ N/A^2. Note that $\vec{r} - \vec{r}\,'$ is the vector displacement from the charge to the field point at $\vec{r}$ and that the direction of $\vec{B}$ is determined by the vector product of $q\vec{v}$ and this displacement. Fields of oppositely charged particles are in opposite directions, all else being the same.

A moving charge exerts a magnetic force on another moving charge. If charge q_1 moves with velocity $\vec{v}_1$ and $\vec{B}_2$ is the magnetic field produced by a second charge q_2, the magnetic force of q_2 on q_1 is given by

$$\vec{F}_1 = q_1 \vec{v}_1 \times \vec{B}_2. \qquad (14\text{-}2)$$

The first charge also exerts a magnetic force on q_2: $\vec{F}_2 = q_2 \vec{v}_2 \times \vec{B}_1$ where $\vec{B}_1$ is the magnetic field produced by q_1.

Fig. 14-1 gives the flow chart for a program which can be used to calculate the total magnetic field produced by a collection of moving charges. To save storage space, the physical situation we deal with is limited to two charges, both moving in the z direction. The first particle has charge q_1, velocity $\vec{v}_1 = v_1 \hat{k}$, and is at the origin at the time for which the field is calculated. The second particle has charge q_2, velocity $\vec{v}_2 = v_2 \hat{k}$, and is at $y'\hat{j} + z'\hat{k}$ at the time for which the field is calculated.

Equations for the field components are derived by substituting the appropriate expressions for the velocities and positions of the charges into Eqn. 14-1 and summing the contributions for the two charges. We find the field at the point x,y in the x,y plane. At this point, the z component of the field vanishes,

$$B_x = -\frac{\mu_0}{4\pi}\left[\frac{q_1 v_1 y}{(x^2+y^2)^{3/2}} + \frac{q_2 v_2 (y-y')}{\left[x^2+(y-y')^2+z'^2\right]^{3/2}}\right] \quad (14\text{-}3)$$

and

$$B_y = \frac{\mu_0}{4\pi}\left[\frac{q_1 v_1 x}{(x^2+y^2)^{3/2}} + \frac{q_2 v_2 x}{\left[x^2+(y-y')^2+z'^2\right]^{3/2}}\right]. \quad (14\text{-}4)$$

Storage allocation for the program is as follows.

    X1 q_1 (at line 100), $(\mu_0/4\pi)q_1 v_1$ (at line 110 and thereafter),
    X2 v_1 (at line 100), $(x^2+y^2)^{3/2}$ (at line 160 and thereafter),
    X3 q_2 (at line 120), $(\mu_0/4\pi)q_2 v_2$ (at line 130 and thereafter),
    X4 v_2 (at line 120), $\left[x^2+(y-y')^2+z'^2\right]^{3/2}$ (at line 160 and thereafter),
    X5   y coordinate of the second charge,
    X6   z coordinate of the second charge,
    X7   x coordinate of field point,
    X8   y coordinate of field point,
    X9   x component of the magnetic field,

and   X10   y component of the magnetic field.

The charge and velocity always appear in the combination $(\mu_0/4\pi)qv$ so this quantity is calculated for each charge immediately after the charge and velocity information is entered, at lines 100 and 120.

Figure 14-1. Flow chart for a program to calculate the magnetic field of two moving charges.

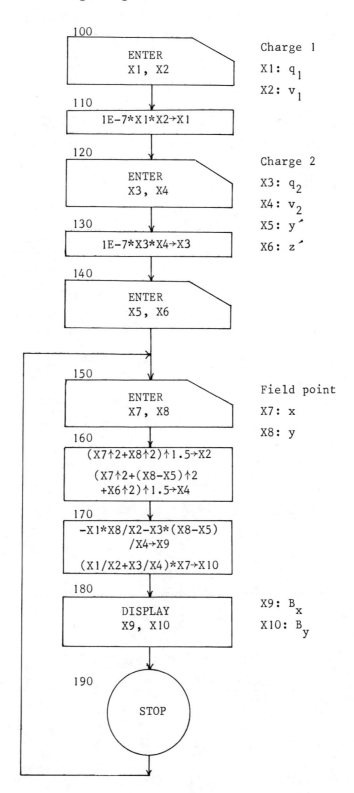

At line 140, the coordinates of the second charge are entered and, at line 150, the coordinates of the field point are entered. The combinations $(x^2+y^2)^{3/2}$ and $[x^2+(y-y')^2+z'^2]^{3/2}$ are evaluated at line 160. The field components are computed at line 170. B_x is stored in X9 and B_y in X10. After the field components are displayed and the machine restarted, it returns to line 150, where it accepts the coordinates of another field point.

Each of the two moving charges exerts magnetic force on the other and the magnetic forces obey Newton's third law. The force of charge 1 on charge 2 is equal in magnitude and opposite in direction to the force of charge 2 on charge 1. If the two charges have the same sign and move in the same direction or, if they have opposite sign and move in opposite directions, they attract each other.

<u>Problem 1.</u> Charge $q_1=6.3\times10^{-9}$ C moves so that its position vector is $\vec{r}_1=3.2\times10^4 t\hat{k}$ as a function of time and charge $q_2=7.1\times10^{-9}$ C moves so that its position vector is $\vec{r}_2=2\times10^{-3}\hat{j}+(1.3\times10^{-3}+4.6\times10^4 t)\hat{k}$ as a function of time. Distances are in meters and time in seconds. Answer the following questions for time $t=0$.

a. What is the magnetic force of q_1 on q_2? Is it one of attraction or one of repulsion? The program of Fig. 14-1 can be used to calculate the field of a single charge by setting the other charge equal to zero. To solve this problem, you must relabel the charges and shift the coordinate system.

b. What is the magnetic force of q_2 on q_1? In what direction does it act? The machine will try to calculate the field at the origin due to q_1 and will have trouble. Use $x=1\times10^{-30}$ m, $y=0$ as the field point instead of the origin.

c. A third charge $q_3=5.8\times10^{-7}$ C moves with position vector $\vec{r}_3=(3.6\times10^{-3}-1.7\times10^4 t)\hat{i}+2.6\times10^4 t\hat{j}$. Distances are in meters and time in seconds. What is the magnetic force on q_3?

d. Answer the questions of part a for $q_1=-6.3\times10^{-9}$ C and all else the same.

e. Answer the questions of part b for $q_1=-6.3\times10^{-9}$ C and all else the same.

<u>Problem 2.</u> Charge $q_1=7.7\times10^{-9}$ C moves along the z axis with velocity $v_1=1.8\times10^4$ m/s in the positive z direction. At $t=0$, it is at the origin. Charge $q_2=9.1\times10^{-9}$ C moves along the line $x=0$, $y=3.8\times10^{-3}$ m with velocity 6.6×10^4 m/s, also in the positive z direction. At $t=0$ its z coordinate is -0.7×10^{-3} m. Answer the following questions for $t=0$. To answer part b, the program must be revised to take into account the z

coordinate of the field point.

a. In the region between the charges, the two magnetic fields tend to cancel. Calculate the magnetic field for the points y=0.5 mm, 1 mm, 1.5 mm, 2 mm, 2.5 mm, 3 mm, and 3.5 mm along the y axis. Use a fit to a fourth order polynomial and the binary search program to find the point on the y axis where the field vanishes.

b. The point you found is part of a line along which the magnetic field vanishes. All points on the line have x=0. Find the y coordinates of the points for which z=1 mm, 0.5 mm, and -0.5 mm. Is the line straight? What happens to it as time goes on?

Magnetic field lines can be plotted using the same strategy as was used to plot electric field lines in section 12.2. If x,y,z is a point on a field line, then

$$x+(B_x/B)\Delta s, \qquad y+(B_y/B)\Delta s, \qquad z+(B_z/B)\Delta s \qquad (14\text{-}5)$$

is another point on the same field line, separated from the first by the distance Δs as measured along the line.

We consider the same situation as previously: charge q_1 at the origin and charge q_2 at $y´,z´$ in the y,z plane, both moving in the z direction. We plot field lines in the x,y plane.

The flow chart is shown in Fig. 14-2. Calculation of the field proceeds as before, except that the factor $\mu_0/4\pi$ is excluded since it appears in both numerator and denominator of B_x/B and B_y/B.

The first eight storage locations are used as in the program of Fig. 14-1. X7 and X8 are the x and y coordinates, respectively, of a point on the field line. Initial values for these coordinates are entered at line 100 and coordinates for other points on the same line are computed by the machine.

X11 contains the distance Δs between successive points along the field line and X12 contains the number of points N to be found before the coordinates are displayed. For high accuracy, it is necessary to pick Δs small, but for plotting purposes, one does not need all the points generated. N can be made large to save time and avoid copying an enormous amount of data.

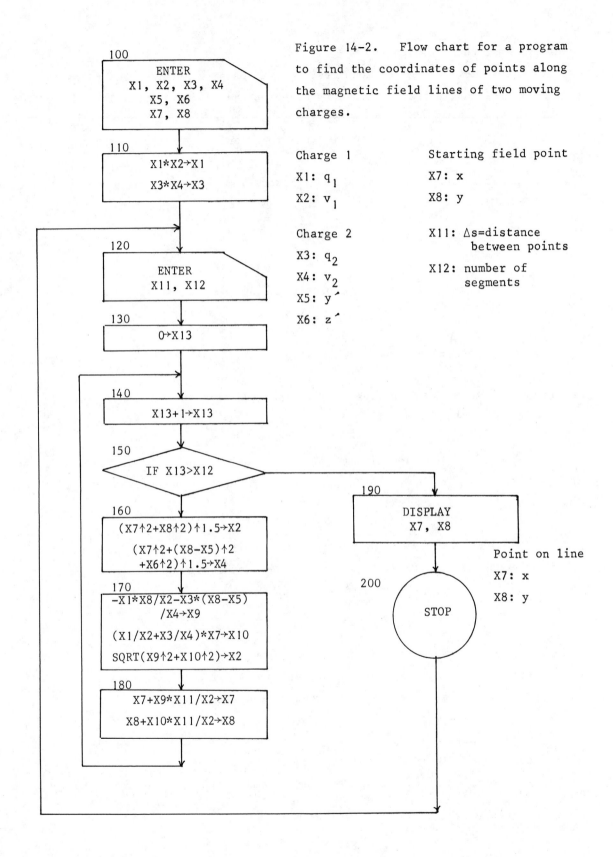

Figure 14-2. Flow chart for a program to find the coordinates of points along the magnetic field lines of two moving charges.

Charge 1
X1: q_1
X2: v_1

Charge 2
X3: q_2
X4: v_2
X5: y'
X6: z'

Starting field point
X7: x
X8: y

X11: Δs=distance between points
X12: number of segments

Point on line
X7: x
X8: y

At line 170, B_x, B_y and $B=(B_x^2+B_y^2)^{\frac{1}{2}}$ are calculated and, at line 180, the coordinates of the next point on the field line are computed. After N points are found, the coordinates of the last one are displayed and, when the machine is restarted, it returns to line 120 to receive new values of Δs and N.

Unlike electric field lines, magnetic field lines do not start and stop at charges but rather form closed contours. For a single moving charge, the magnetic field lines are circular. Any chosen field line lies in a plane perpendicular to the velocity vector for the particle and the line of the velocity vector goes through the center of the circle formed by the line. These circular field lines exist in front of and behind the particle as well as in the plane containing the particle.

<u>Problem 3.</u> A charge $q=3\times10^{-9}$ C moves along the z axis with velocity 5.5×10^5 m/s in the positive z direction.

a. Suppose the charge is at $z=-0.5$ m and plot the field line through $x=0.25$ m, $y=0$, $z=0$. The graph should show the x,y plane. The program of Fig. 14-2 can be used if q_1 is set equal to zero. Use $\Delta s=2.5\times10^{-3}$ m and plot every 10 points for a little more than one fourth of the circle. This is a field line in a plane ahead of the charge.

b. Suppose the charge is at $z=0$ and plot the field line through $x=0.25$ m, $y=0$. $z=0$. This is a field line in the plane of the charge.

c. Use the program of Fig. 14-1 to find the components and magnitude of the field near the point on the line of part b with $x=0.14$ m. Graphically show that the field is tangent to the line through the field point. To see if the magnitude of the field is constant on the line, also calculate it for $x=0.25$ m, $y=0$, $z=0$.

When there is more than one charge, the magnetic field lines no longer form circles and the magnitude of the field is no longer constant along a line. The following two problems demonstrate these statements.

<u>Problem 4.</u> Two identical charges, one on the line $x=0$, $y=0$ and the other on the line $x=0$, $y=0.03$ m, move side by side in the positive z direction with speed 3.7×10^5 m/s. Take each charge to be 4.6×10^{-9} C and plot the following field lines, in the x,y plane, for the time when the charges are at the x,y plane.

a. The field line through x=0, y=-0.7×10^{-2} m. Use $\Delta s = 1.5 \times 10^{-4}$ m and plot every 20 points.

b. The field line through x=0, y=-1×10^{-2} m. Use $\Delta s = 2 \times 10^{-4}$ m and plot every 20 points.

c. Calculate the components and magnitude of the magnetic field near the points where these lines cross the x axis and near the points, on the lines, where y=0.7×10^{-2} m. Note that the magnitude is not constant on a line. Graphically show that the field is tangent to the line through the field point.

d. Explain the shape of the line of part b in terms of the fields of the individual charges.

Problem 5. Repeat the calculation of problem 4 with $q_2 = -4.6 \times 10^{-9}$ C and all else the same.

There is an error of about 2×10^{-4} m in the calculated coordinates of the last points found on each line. If you wish to decrease the error, reduce Δs. The running time increases proportionally.

14.2 Magnetic Fields of Circular Current Loops

We consider the magnetic field produced by a current, a collection of many moving charges, distributed so densely that it is impractical to sum the individual fields, one at a time. We assume the charge moves along a wire so thin that we may neglect its cross section compared to the distance from the wire to the field point.

We divide the wire into many short segments and use $\Delta \vec{\ell}$ to represent a vector in the direction of the current, with length equal to that of a segment. If a segment center is at $\vec{r}'$ and the wire carries current I, then the field it produces at $\vec{r}$ is

$$\vec{B}(\vec{r}) = \frac{\mu_0}{4\pi} I \frac{\Delta \vec{\ell} \times (\vec{r} - \vec{r}')}{|\vec{r} - \vec{r}'|^3} . \qquad (14\text{-}6)$$

To obtain the total field, we sum the contributions of the segments. In the limit $\Delta \vec{\ell} \to 0$, the sum becomes an integral and the result is written

$$\vec{B}(\vec{r}) = \frac{\mu_0}{4\pi} I \int \frac{d\vec{\ell} \times (\vec{r} - \vec{r}')}{|\vec{r} - \vec{r}'|^3} . \qquad (14\text{-}7)$$

We consider current elements which lie in the x,y plane and, consequently, take $\vec{r}\,' = x'\hat{i} + y'\hat{j}$. If $\vec{r} = x\hat{i} + y\hat{j} + z\hat{k}$ and $d\vec{\ell} = d\ell_x \hat{i} + d\ell_y \hat{j}$, then the cartesian components of Eqn. 14-7 are

$$B_x(\vec{r}) = \frac{\mu_0 I}{4\pi} \int \frac{z\, d\ell_y}{\left[(x-x')^2 + (y-y')^2 + z^2\right]^{3/2}}, \qquad (14\text{-}8)$$

$$B_y(\vec{r}) = -\frac{\mu_0 I}{4\pi} \int \frac{z\, d\ell_x}{\left[(x-x')^2 + (y-y')^2 + z^2\right]^{3/2}}, \qquad (14\text{-}9)$$

and
$$B_z(\vec{r}) = \frac{\mu_0 I}{4\pi} \frac{(y-y')\, d\ell_x - (x-x')\, d\ell_y}{\left[(x-x')^2 + (y-y')^2 + z^2\right]^{3/2}}. \qquad (14\text{-}10)$$

The integration program of Fig. 8-1 can be used to add contributions from the various segments. In order to carry out this program, it is necessary to specify the circuit as a whole and $d\vec{\ell}$ for each segment. We demonstrate by calculating the field of a circular circuit.

We suppose current flows around a circle of radius R, in the x,y plane and centered at the origin, as shown in Fig. 14-3.

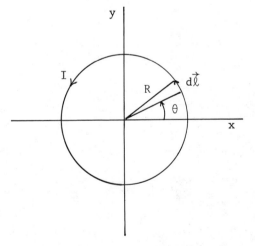

Figure 14-3. A circular loop of wire carrying current I.

We use θ as the variable of integration. The position of a point on the circle is given by

$$\vec{r}\,' = R\cos\theta\, \hat{i} + R\sin\theta\, \hat{j} \qquad (14\text{-}11)$$

and
$$d\vec{\ell} = (-R\sin\theta\, \hat{i} + R\cos\theta\, \hat{j})\, d\theta. \qquad (14\text{-}12)$$

Now

$$B_x(\vec{r}) = \frac{\mu_0 I}{4\pi} Rz \int_0^{2\pi} \frac{\cos\theta \, d\theta}{\left[(x-R\cos\theta)^2+(y-R\sin\theta)^2+z^2\right]^{3/2}} \,, \qquad (14\text{-}13)$$

$$B_y(\vec{r}) = \frac{\mu_0 I}{4\pi} Rz \int_0^{2\pi} \frac{\sin\theta \, d\theta}{\left[(x-R\cos\theta)^2+(y-R\sin\theta)^2+z^2\right]^{3/2}} \,, \qquad (14\text{-}14)$$

and

$$B_z(\vec{r}) = -\frac{\mu_0 I}{4\pi} R \int_0^{2\pi} \frac{\left[\sin\theta(y-R\sin\theta)+\cos\theta(x-R\cos\theta)\right] d\theta}{\left[(x-R\cos\theta)^2+(y-R\sin\theta)^2+z^2\right]^{3/2}} \,. \qquad (14\text{-}15)$$

Each of these integrals can be evaluated using the program of Fig. 8-1. We give the program for a slightly special case. We take I=1 A and calculate the magnetic field at a point in the y,z plane. For other currents, simply multiply the field obtained for 1 A by the actual current. For x≠0, the field can be found using a symmetry argument and the field at a point with the same y and z coordinates but with x=0.

Eqns. 14-13, 14-14, and 14-15 become

$$B_x = 0, \qquad (14\text{-}16)$$

$$B_y = \frac{\mu_0}{4\pi} \frac{1}{R} \frac{z}{R} \int_0^{2\pi} \frac{\sin\theta \, d\theta}{\left[1+(y/R)^2+(z/R)^2-2(y/R)\sin\theta\right]^{3/2}} \,, \qquad (14\text{-}17)$$

$$B_z = -\frac{\mu_0}{4\pi} \frac{1}{R} \int_0^{2\pi} \frac{\left[(y/R)\sin\theta - 1\right] d\theta}{\left[1+(y/R)^2+(z/R)^2-2(y/R)\sin\theta\right]^{3/2}} \,, \qquad (14\text{-}18)$$

respectively. For ease in calculation, we have chosen to measure distances in units of R. Thus y/R and z/R appear. We have also used $\sin^2\theta = \cos^2\theta = 1$ to combine some of the terms.

Both of the last two equations have the form

$$B_i = \frac{\mu_0}{4\pi R} \int_0^{2\pi} f(\theta) \, d\theta \,. \qquad (14\text{-}19)$$

We follow the integration scheme of Fig. 8-1 but choose to evaluate both integrals in the same program. The interval from 0 to 2π is divided into N equal segments and values of the integrand at successive end points are labelled f_0, f_1, f_2, ... f_N, where, in the present case, f_0 is the function at $\theta=0$ and f_N is the function at $\theta=2\pi$. The integral is then

$$(\Delta\theta/3) \left[(f_0-f_N)+2(f_2+f_4+f_6+ \ldots f_N) + 4(f_1+f_3+ \ldots)\right]$$
$$= (\Delta\theta/3)\left[2(f_2+f_4+f_6+ \ldots f_N) + 4(f_1+f_3+ \ldots)\right]. \qquad (14\text{-}20)$$

The last form follows since $f_0=f_N$. N must be an even number.

The flow chart is shown in Fig. 14-4. The meanings of the storage location symbols are as follows.

    X1   radius of circuit,
    X2   number of segments,
    X3   y coordinate of field point, y/R (line 120 and thereafter),
    X4   z coordinate of field point, z/R (line 120 and thereafter),
    X5   angle θ,
    X6   used to collect odd labelled terms for B_y,
    X7   used to collect even labelled terms for B_y,
    X8   used to collect odd labelled terms for B_z,
    X9   used to collect even labelled terms for B_z,
    X10   the combination $1+(y/R)^2+(z/R)^2$,
    X11 $\sin\theta$,
and   X12   the combination $\left[1+(y/R)^2+(z/R)^2-2(y/R)\sin\theta\right]^{3/2}$.

Immediately upon entry, at line 120, y and z are divided by R. The combination $1+(y/R)^2+(z/R)^2$ is also calculated and stored in X10. The factors $(\mu_0/4\pi)(1/R)(z/R)$ in the equation for B_y and $(\mu_0/4\pi)(1/R)$ in the equation for B_z are not included in the calculation until after the integrals have been evaluated. Multiplication by these factors is accomplished at line 200. For each value of θ, $\sin\theta$ is calculated once and stored in X11. It is used in the three following computations, at lines 170 and 190. Rather than count the number of times around the loop, the angle θ is checked, at line 160, to see if it has reached 2π. If it has, all intervals have been considered and the machine goes to line 200 where the contributions to the integrals are assembled and

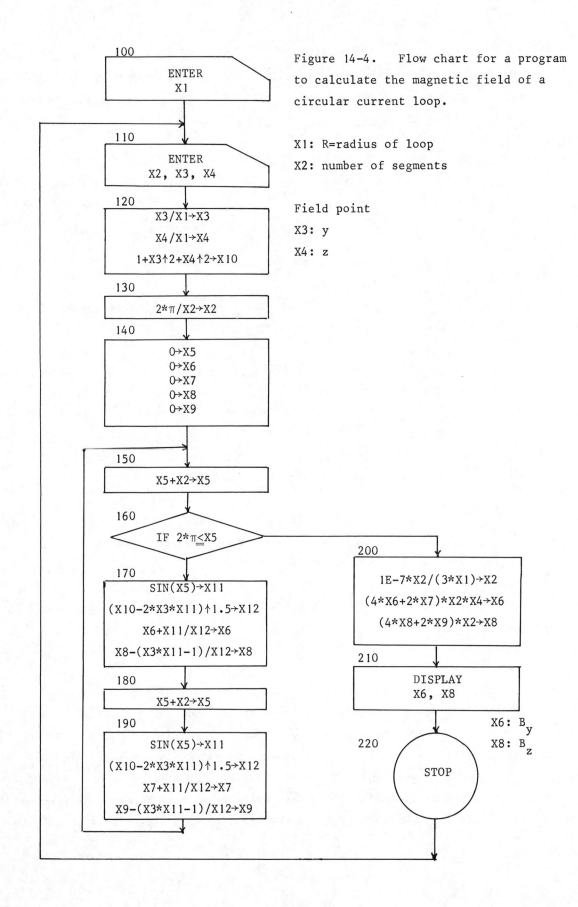

Figure 14-4. Flow chart for a program to calculate the magnetic field of a circular current loop.

X1: R=radius of loop
X2: number of segments

Field point
X3: y
X4: z

X6: B_y
X8: B_z

the results multiplied by the appropriate factors. At line 210, B_y is in X6 and B_z is in X8. In all other respects, the program is a straight forward adaptation of the program of Fig. 8-1.

The following problems provide some situations for which the program can be used.

<u>Problem 1.</u> If the field point is on the x axis, Eqns. 14-17 and 14-18 can be evaluated in closed form. For such a field point $B_x=0$, $B_y=0$, and

$$B_z(z) = \frac{\mu_0 I R^2}{2(R^2+z^2)^{3/2}}.$$

a. Use the above equation to calculate the magnetic field at the center of a 1 m radius circular loop carrying 1 A of current.
b. Use N=2 and the program of Fig. 14-4 to calculate the field at the center of a 1 m radius circular loop carrying 1 A of current. Since the integrand is constant, the program should give results which are exact except for round off error.

We wish to find the appropriate number of segments to use when we use the program of Fig. 14-4 to find the magnetic field.

<u>Problem 2.</u> Consider a 1 m radius circular loop in the x,y plane, centered at the origin. It carries 1 A of current.
a. Find the magnetic field at the point x=0, y=0.5 m, z=0. Start with N=2 and, on the next trial, double N. Continue to double N until results of two successive trials agree to 3 significant figures.
b. Repeat the calculation of part a for the following points along the x=0, y=0.5 m line: z=0.1 m, 1 m, and 10 m. For each point, note the value of N which gives 3 significant figure accuracy.

Helmholtz coils consist of two loops, parallel to each other, with the same axis, and with current of the same magnitude, flowing in the same direction. In the region between the coils, the two magnetic fields tend to be in roughly the same direction and, when the distance between the coils is equal to the radius of one of them, the field in the region between is particularly uniform. For this reason, Helmholtz coils are often used to produce magnetic fields in the laboratory. With the program of Fig. 14-4, we can investigate the uniformity of the magnetic field in the region between the coils.

Problem 3. Two circular loops of wire, each with radius 1 m, are placed parallel to the x,y plane, with their centers on the z axis. The center of one is at z=0 while the center of the other is at z=2 m. A current of 1 A flows in both of the loops, in a counterclockwise direction when viewed from the positive z axis.

a. Calculate the magnetic field with 3 figure accuracy for field points at z=0.6 m, 0.8 m, and 1 m, all on the z axis. The last point is at the center of the region between the two loops and the other points are 0.2 m and 0.4 m away, respectively. For each field point, you will need to run the program twice, once for each coil. The first field point, z=0.6 m, is 0.6 m from the center of the first coil and 1.4 m from the center of the second coil. The program is run with X4=0.6, then with X4=-1.4, and the two fields are vectorially summed.

We use these calculated fields to test how homogeneous the field is along the z axis between the loops. We can define a quantity which is a measure of the homogeneity of the field. For each of the first two points, subtract the magnitude of the field at the center from the magnitude of the field at the point and divide the result by the magnitude of the field at the center. If we denote this measure of homogeneity by h, then

$$h(z) = \frac{|B(z) - B_c|}{|B_c|}$$

where B_c is the magnitude of the field at the center of the region.

If h is small, the field does not change much with position and is said to be homogeneous. If, on the other hand, h is large, the field is said to be inhomogeneous. For good laboratory magnets, h may be on the order of 10^{-6} or less over several centimeters. The distance between the poles is, of course, usually smaller than 2 m.

b. Calculate h for the first two field points of part a.

c. Now consider the same loops, but separated by 1 m, the radius of one of them. The first loop still has its center at the origin, but now the center of the second loop is at z=1 m on the z axis. Calculate the magnetic field at z=0.1 m, 0.3 m, and 0.5 m, all on the z axis. The last point is at the center of the region between the loops and the other points are 0.2 m and 0.4 m away respectively, as before.

d. For the first two points of part c, calculate h. Has the field become more or less homogeneous?

We can also investigate homogeneity in a transverse direction.

Problem 4. Consider the two loops of problem 3, separated by 2 m.
a. Calculate the magnetic field for the following points along the line x=0, z=1 m: y=0, 0.2 m, and 0.4 m.
b. The field may have two non-vanishing components, B_y and B_z, and we define a measure of homogeneity for each component. For the z component

$$h_z = \frac{|B_z(y) - B_z(y=0)|}{|B_z(y=0)|}$$

and, for the y component,

$$h_y = \frac{|B_y(y)|}{|B_z(y=0)|}$$

Both are zero for a perfectly homogeneous field.

Evaluate h_z and h_y for the last two points of part a.

c. Change the separation of the loops to 1 m by placing the center of the second loop at z=1 m on the z axis. Calculate the magnetic field for the following points on the line x=0, z=0.5 m: y=0, 0.2 m, and 0.4 m.

d. For the first two points of part c, evaluate h_z and h_y. Has the field become more or less homogeneous when the loops are moved closer together?

If the loops are moved still closer so that their separation is less than the radius, the field becomes less homogeneous.

A solenoid consists of a wire wrapped around the outside of a cylinder. It can be approximated by a series of loops, parallel to each other and centered on the same axis. If there is a large number of loops (a solenoid of infinite length) and if the thickness of the wire can be neglected, the magnetic field inside the cylinder is nearly uniform, nearly parallel to the axis, and has the magnitude

$$B = \mu_0 I n$$

where n is the number of loops per unit length of the solenoid. The field outside such a solenoid is zero.

The field, of course, is produced by the current flowing in the loops and is the superposition of the fields produced by the individual loops. It is inconvenient to consider too large a number of loops, but the following calculations, involving a small number of loops, should convince you that the field becomes more nearly uniform inside and more nearly zero outside as the number of loops is increased.

Problem 5. Consider a solenoid formed by 11 circular loops, each of radius 0.03 m, placed with centers on the z axis, from z=0, to z=0.1 m at 0.01 m intervals. The loops are parallel to the x,y plane and each carries a current of 1 A, all in the same direction.

a. Calculate the magnetic field at z=0.05 m on the z axis. Obtain 3 significant figure accuracy. Calculate the ratio of the z component of the field to the value of $\mu_0 I n$.
b. Calculate the magnetic field at z=0.05 m on the line x=0, y=0.02 m. Calculate the ratio of the z component to the value of $\mu_0 I n$.
c. Calculate the magnetic field at z=0.05 m on the line x=0, y=0.06 m. Calculate the ratio of the z component to the value of $\mu_0 I n$.

It is instructive to work problem 5 partially by hand. For a given field point, use the program of Fig. 14-4 to calculate separately the contribution of each loop. Then add the contributions. Note that the total is far more uniform than the field of any single loop.

With more loops, the field is closer to the value $\mu_0 In$ in the interior of the solenoid. For the field points we have chosen, it is more important to add more loops at the same spacing than it is to decrease the spacing between the loops. We therefore consider a longer solenoid in the next problem.

Problem 6. Consider a solenoid consisting of 21 loops, centered every 0.01 m. along the z axis from z=-0.05 m to z=0.15 m. Each carries 1 A of current, all in the same direction.
a. Find the magnectic field at z=0.05 m on the z axis. Compare the result with $\mu_0 In$ and with the appropriate answer to part a of problem 5.
b. Find the magnetic field at x=0, y=0.06 m, z=0.05 m. Compare the result with $\mu_0 In$ and with the appropriate answer to part c of problem 5.

14.3 Ampere's Law

Ampere's law deals with any surface bounded by a closed contour. The current flowing through the surface, from one side to the other, is related to the line integral of the magnetic field around the perimeter. The relationship is

$$\oint \vec{B} \cdot d\vec{\ell} = \mu_0 I \tag{14-21}$$

where I is the net current through the surface. The net current is computed as the algebraic sum of all currents through the surface, currents in different directions being assigned different signs. The law is different if there is a time dependent electric field present and we assume there is none.

We consider a piece of surface in the shape of a square with edge length a, in the x,y plane, and centered at the origin. See Fig. 14-5. A current carrying wire is parallel to the z axis and we take current flowing in the positive z direction to be positive. Then, by the convention associated with Ampere's law, $d\vec{\ell}$ points in the counterclockwise direction when viewed from the positive z axis.

We also assume the current carrying wire has infinite length. Then

$$B = \frac{\mu_0 I}{2\pi R} \tag{14-22}$$

gives the magnitude of the field a distance R from the wire. $\vec{B}$ is tangent to the circle of radius R, centered on the wire.

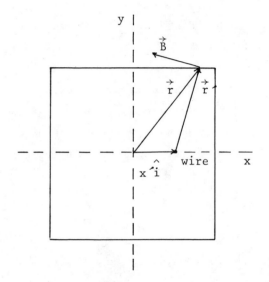

Figure 14-5. A square contour used to calculate the line integral of the magnetic field. The z axis is out of the page.

The current carrying wire cuts the plane of the square at $x'\hat{i}$ on the x axis and the field point is at $\vec{r}=x\hat{i}+y\hat{j}$ on the boundary of the square. The vector from the wire to the field point is labelled $\vec{r}\,'$:

$$\vec{r}\,' = \vec{r} - x'\hat{i} = (x-x')\hat{i}+y\hat{j} \tag{14-23}$$

and
$$|\vec{r}\,'| = R = \left[(x-x')^2+y^2\right]^{\frac{1}{2}} \tag{14-24}$$

The magnetic field is in the direction $\hat{k}\times\vec{r}\,'$ so

$$\vec{B} = \frac{\mu_0 I}{2\pi}\frac{\hat{k}\times\vec{r}\,'}{|\vec{r}\,'|^2} = -\frac{\mu_0 I}{2\pi}\frac{y\hat{i}-(x-x')\hat{j}}{(x-x')^2+y^2} . \tag{14-25}$$

We consider separately the four sides of the square. Across the top $y=a/2$, $\vec{d\ell}=dx\hat{i}$, and

$$\int \vec{B}\cdot\vec{d\ell} = -\frac{\mu_0 I}{2\pi}\frac{a}{2}\int_{a/2}^{-a/2}\frac{dx}{(x-x')^2+(a/2)^2} . \tag{14-26}$$

We have used the limits of integration to specify the direction of $d\vec{\ell}$. Since the lower limit is greater than the upper limit, dx is negative and $d\vec{\ell}$ points from right to left across the top, as it should.

Down the left side, $x=-a/2$, $d\vec{\ell}=dy\hat{j}$, and

$$\int \vec{B} \cdot d\vec{\ell} = -\frac{\mu_0 I}{2\pi} \left(\frac{a}{2} + x'\right) \int_{a/2}^{-a/2} \frac{dy}{(x'+a/2)^2 + y^2} . \quad (14\text{-}27)$$

Across the bottom, $y=-a/2$, $d\vec{\ell}=dx\hat{i}$, and

$$\int \vec{B} \cdot d\vec{\ell} = \frac{\mu_0 I}{2\pi} \frac{a}{2} \int_{-a/2}^{a/2} \frac{dx}{(x-x')^2 + (a/2)^2} . \quad (14\text{-}28)$$

Up the right side, $x=a/2$, $d\vec{\ell}=dy\hat{j}$, and

$$\int \vec{B} \cdot d\vec{\ell} = \frac{\mu_0 I}{2\pi} \left(\frac{a}{2} - x'\right) \int_{-a/2}^{a/2} \frac{dy}{(-x'+a/2)^2 + y^2} . \quad (14\text{-}29)$$

Because the wire cuts the plane of the square on the x axis, the contributions of the top and bottom are the same. The signs in front are different but the limits of integration are reversed. The integral around the entire square is the sum of the individual contributions of the sides:

$$\oint \vec{B} \cdot d\vec{\ell} = \frac{\mu_0 I}{2\pi} \left[a \int_{-a/2}^{a/2} \frac{dx}{(x-x')^2 + (a/2)^2} + \left(\frac{a}{2} + x'\right) \int_{-a/2}^{a/2} \frac{dy}{(x'+a/2)^2 + y^2} \right.$$

$$\left. + \left(\frac{a}{2} - x'\right) \int_{-a/2}^{a/2} \frac{dy}{(-x'+a/2)^2 + y^2} \right] . \quad (14\text{-}30)$$

Here we have inverted the integration limits and changed the sign in front of the second integral.

The program of Fig. 8-1 can be used to evaluate the integrals. In each case $-a/2$ is entered into X1 and $+a/2$ is entered into X2. Each side of the square is divided into N segments and this number is entered into X3. It must be even.

For the first integral, the function used at lines 130 and 170 is

$$1/((X1-x')^2+(a/2)^2)$$

where the user must substitute appropriate numbers for x' and a. When the integration has been completed, the result must be multiplied by $\mu_0 Ia/2\pi$. This multiplication can be done separately or incorporated into the program, after line 200.

For the second integral, the function used at lines 130 and 170 is

$$1/((a/2 + x')^2+X1^2)$$

and the value of the integral must be multiplied by $(\mu_0 I/2\pi)(x'+a/2)$. For the third integral, the function is

$$1/((a/2 - x')^2+X1^2)$$

and the value of the integral must be multiplied by $(\mu_0 I/2\pi)(a/2-x')$.

We now use the integration program to investigate some statements about Ampere's law. For the same current through the surface, the value of the integral is the same regardless of the position of the wire. It must pass through the surface, of course. The integral is zero if no current passes through the surface.

Problem 1. Suppose 1 A flows in the positive z direction. The square described above lies in the x,y plane. Take a=2 m.
a. Evaluate $\oint \vec{B} \cdot d\vec{\ell}$ for the following positions of the wire: $x'=0$, 0.4 m, 0.8 m, and 1.2 m. The first three points are inside the square while the last is outside. Use N=64.
b. For each of the current configurations of part a, evaluate $\mu_0 I$ where I is the current passing through the square. Compare results with the answers to part a.

The current I which appears on the right side of Eqn. 14-21 is the net current through the surface. If two wires pass through, with current flowing in the same direction, the currents add to produce the net current. If the currents are in

Problem 2. For the square described in problem 1:

a. Evaluate the integrals in Eqn. 14-30 for a current of 2 A flowing in the positive z direction at $x'=0$.

b. Evaluate the integrals for a current of 1 A flowing in the positive z direction at $x'=0.5$ m.

c. Evaluate the integrals for a current of 3 A flowing in the positive z direction at any value of x' within the square. Compare the answer to the sum of the answers to parts a and b.

Problem 3. For the square described above:

a. Evaluate the integrals in Eqn. 14-30 for a current of 1 A flowing in the negative z direction at $x'=0.5$ m.

b. Evaluate the integrals for a current of 1 A flowing in the positive z direction at any value of x' within the square. Compare the answer to the sum of the answers to part a of problem 2 and part a of this problem.

Chapter 15
MOTION IN ELECTRIC AND MAGNETIC FIELDS

In this chapter, we examine the motion of charged particles in electric and magnetic fields and magnetic forces and torques on current carrying wires. This material supplements topics covered in Chapters 27 and 33 of PHYSICS, Chapters 24 and 30 of FUNDAMENTALS OF PHYSICS, and similar sections of other texts. Use of the programmable calculator allows us to follow the motion of two charges as they interact, to follow a charge as it moves among other charges, and to follow the somewhat complicated motion of a charge in a non-uniform magnetic field. We also calculate the force on square and circular loops of wire in non-uniform magnetic fields.

15.1 The Force Equations

The force on a moving charge q in an electric field $\vec{E}$ and a magnetic field $\vec{B}$ is given by

$$\vec{F} = q\vec{E} + q\vec{v}\times\vec{B} \qquad (15\text{-}1)$$

where $\vec{v}$ is the velocity of the charge. The fields are, of course, evaluated at the position of the charge. If the fields are different at different places, the charge experiences a force which changes as it moves. The magnetic force also changes as the velocity of the charge changes.

We deal with two dimensional problems for which the velocity and the force are both in the x,y plane. In particular, the electric fields have no z components and the magnetic fields have no x or y components. For these problems, the program of Fig. 7-1 can be used to find the subsequent motion of the charge, given its initial position and velocity.

The x and y components of the acceleration are given by

$$a_x = F_x/m = (q/m)(E_x + v_y B_z) \qquad (15\text{-}2)$$

and
$$a_y = F_y/m = (q/m)(E_y - v_x B_z) \, , \qquad (15\text{-}3)$$

respectively. Here m is the mass of the particle. Note that, because of the cross product nature of the magnetic force, the x component of the velocity enters the y component of the acceleration and the y component of the velocity enters the x component of the acceleration. Eqn. 15-2 should be used at line 150 of the program and Eqn. 15-3 should be used at line 160. For each integration interval, the components of the fields are evaluated using the position and velocity of the charge at the beginning of the interval.

If $\vec{B}=0$ and only an electric field acts on the particle, the force is independent of the velocity and the program of Fig. 7-3 can be used. Because of the greater speed of this program, compared to the program of Fig. 7-1, it is highly recommended that it be used whenever possible. The function f_x, which is evaluated at lines 120, 170, and 200 is then $(q/m)E_x$ and the function f_y, which is evaluated at lines 130, 180, and 210, is $(q/m)E_y$. In each case, the values of the field components are those appropriate for the position of the charge.

For both programs, X1 and X2 contain the x and y coordinates, respectively, of the charge's position and X3 and X4 contain the x and y components, respectively, of its velocity. X6 is the time interval over which the force is assumed constant and X7 is the number of intervals considered by the machine before results are displayed.

The fundamental problem is to find the positions and velocities as functions of time for two charged particles which move under the influence of each other. We can get an idea of the relative magnitudes of the electric and magnetic forces by considering a special but not untypical situation: two identical particles.

The particles are placed equal distances from the x axis, one on each side and they start travelling with identical velocities in the x direction. The force of the first on the second is equal in magnitude and opposite in direction to the force of the second on the first and their trajectories remain symmetric about the x axis as shown in Fig. 15-1.

If the first particle is at $\vec{r}_1=x\hat{i}+y\hat{j}$ and it has velocity $\vec{v}_1=v_x\hat{i}+v_y\hat{j}$, then the second particle is at $\vec{r}_2=x\hat{i}-y\hat{j}$ and it has velocity $\vec{v}_2=v_x\hat{i}-v_y\hat{j}$. We calculate the force on particle 1. The electric and magnetic fields at $\vec{r}_1$, due to particle 2 are given by

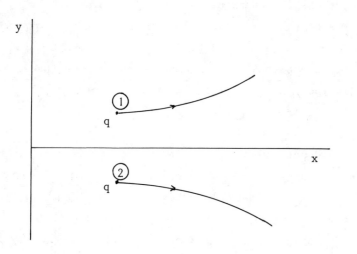

Figure 15-1. Two identical charges moving in the x,y plane.

$$\vec{E}(\vec{r}_1) = \frac{q}{4\pi\epsilon_0} \frac{\vec{r}_1-\vec{r}_2}{|\vec{r}_1-\vec{r}_2|^3} = \frac{q}{4\pi\epsilon_0} \frac{1}{4y^2} \hat{j}, \quad (15\text{-}4)$$

and

$$\vec{B}(\vec{r}_1) = \frac{\mu_0 q}{4\pi} \frac{\vec{v}_2 \times (\vec{r}_1-\vec{r}_2)}{|\vec{r}_1-\vec{r}_2|^3} = \frac{\mu_0 q}{4\pi} \frac{v_x}{4y^2} \hat{k}, \quad (15\text{-}5)$$

respectively. The components of the force on the first particle can be written

$$F_x = \frac{q^2}{4y^2} \frac{\mu_0}{4\pi} v_x v_y \quad (15\text{-}6)$$

and

$$F_y = \frac{q^2}{4y^2} \left(\frac{1}{4\pi\epsilon_0} - \frac{\mu_0}{4\pi} v_x^2 \right) \quad (15\text{-}7)$$

respectively. In SI units

$$F_x = \frac{q^2}{4y^2} v_x v_y \times 10^{-7} \quad (15\text{-}8)$$

and

$$F_y = \frac{q^2}{4y^2} (8.98755 \times 10^9 - v_x^2 \times 10^{-7}). \quad (15\text{-}9)$$

The magnetic force is always less than the electric force. At low speeds, the magnetic force is negligible, but it becomes comparable to the electric force as the velocity components near the speed of light. To obtain this result, equate the two terms which appear in the parentheses of Eqn. 15-9 and solve for v_x.

There are many cases for which the electric field is small or vanishes and the magnetic field dominates or produces a force which is comparable to that of the electric field. This can occur when the magnetic field is produced by current in wires, a large laboratory electromagnet, for example. Then, of course, the electric and magnetic fields considered are produced by different sources.

15.2 Motion in an Electric Field

We consider the influence of one charge on the motion of another and start with the symmetric situation described in the previous section.

Problem 1. Two identical particles, each with charge $q=1.9\times10^{-9}$ C and mass $m=6.1\times10^{-15}$ kg, start with velocity $\vec{v}=3\times10^4 \hat{i}$ m/s. Initially, one is at $x=0$, $y=6.7\times10^{-3}$ m and the other is at $x=0$, $y=-6.7\times10^{-3}$ m. Both are in the x,y plane and continue to move in that plane. Consider only the electric forces the particles exert on each other.

a. Plot the trajectories of the particles. Notice that you need integrate Newton's second law for only one of the particles since the position of the other is known, at each instant of time, in terms of the position of the first. Use the program of Fig. 7-3, with $\Delta t=1\times10^{-8}$ s, and plot the positions every 1×10^{-7} s from $t=0$ to $t=1\times10^{-6}$ s.

b. Suppose now that one of the particles has charge $q=-1.9\times10^{-9}$ C, but all other conditions are the same. Plot the trajectories of the particles from $t=0$ to $t=5\times10^{-7}$ s. The force is now one of attraction.

Save the numerical values of the velocity components for the various times considered. They will be used to answer some questions in a later problem.

There is another physical situation for which the solution to part b applies. For certain conditions, the motion of one of the charges of part b is the same as the motion of a charge placed outside the surface of an ideal conductor. If a charge is placed outside a conductor, charge inside is rearranged so that the electric potential is constant throughout the interior of the conductor. We suppose the conductor lies in the region with negative x and that the x,y plane is its surface. We further suppose the metal is grounded so that its potential is the same as the potential at infinity. If the exterior charge is positive, the rearrangement of charge within the conductor means that negative charge appears on the surface and the cumulative effect of this negative charge is to produce an electric field which, in the exterior region, is identical to the field of a single charge, opposite in sign to the exterior charge, and located at the symmetric point as described in the last section. The charge in the exterior region moves under the influence of this so-called image charge and its trajectory is exactly the same as the trajectory plotted in part b of problem 1. At high speeds, magnetic fields must be taken into account, the metal is no longer a region of constant potential, and the correspondence between the two particle system and the problem of the charge outside the conductor can no longer be made.

The program of Fig. 7-3 is useful when the source of the electric field is stationary. We first consider the motion of a charged particle in a field similar to the one plotted in problem 1 of section 12.2. You should recall roughly the disposition of field lines for that problem. A charged particle in a field does not follow the field lines, but rather, the field lines give the direction of the acceleration for the particle. The following problem provides a demonstration.

Problem 2. An electric field is created by charge $q_1 = 7.1 \times 10^{-8}$ C on the y axis at $y = 1 \times 10^{-4}$ m and charge $q_2 = -7.1 \times 10^{-8}$ C on the y axis at $y = -1 \times 10^{-4}$ m. A third charge $q_3 = 3.5 \times 10^{-8}$ C and mass $m = 4.5 \times 10^{-5}$ kg moves in the x,y plane.

a. Show that the components of the force divided by the mass for the third particle are given by

$$a_x = \frac{q_1 q_3}{4\pi\epsilon_0 m} \left[\frac{x_3}{\left[x_3^2 + (y_3 - y_1)^2\right]^{3/2}} - \frac{x_3}{\left[x_3^2 + (y_3 - y_2)^2\right]^{3/2}} \right]$$

and

$$a_y = \frac{q_1 q_3}{4\pi\epsilon_0 m} \left[\frac{y_3 - y_1}{\left[x_3^2 + (y_3 - y_1)^2\right]^{3/2}} - \frac{y_3 - y_2}{\left[x_3^2 + (y_3 - y_2)^2\right]^{3/2}} \right].$$

b. Suppose the third particle is fired toward the origin with speed 2.7×10^3 m/s along the line x=y with both x and y positive. Take its initial position to be 0.3 m from the origin and plot its position every 1×10^{-5} s from t=0 to t=1×10^{-4} s. Use $\Delta t = 5 \times 10^{-7}$ s.

c. Suppose the third particle is fired toward the origin with speed 2.7×10^3 m/s along the positive x axis. Take its initial position to be 0.3 m from the origin and plot its position every 1×10^{-5} s from t=0 to $t=8 \times 10^{-5}$ s. This line cannot be continued unless Δt is made much smaller and the running time increased significantly.

The electrostatic field is conservative. For the two particles of problem 1 and for the single particle of problem 2, the total mechanical energy, the sum of the kinetic and potential energies, is conserved.

Problem 3. For the system of two charges, as described in part a of problem 1, calculate the total mechanical energy as follows.

a. Show that the potential energy is given by

$$U = \frac{q^2}{4\pi\epsilon_0} \frac{1}{2|y|} .$$

b. Evaluate the potential energy, the kinetic energy, and the total energy for the times t=0, 5×10^{-7} s, and 1×10^{-6} s. Note that, as the separation between the charges increases, the potential energy decreases and the kinetic energy increases.

c. Repeat the calculation of part b for the charges of problem 1, part b. Take t=0 and 5×10^{-7} s. The potential energy is now

$$U = - \frac{q^2}{4\pi\epsilon_0} \frac{1}{2|y|}$$

since the charges have opposite signs. As the separation between the charges decreases, the potential energy also decreases (becomes a larger negative number) and the kinetic energy increases.

d. Do the results of parts b and c indicate conservation of energy to within computational error? To obtain better agreement, decrease Δt when solving problem 1.

15.3 Motion in a Magnetic Field

In the absence of an electric field, the x and y components of the acceleration of a particle with charge q, mass m, and velocity $\vec{v}$, in a magnetic field $\vec{B} = B_z \hat{k}$ are given by

$$a_x = (q/m)v_y B_z \qquad (15\text{-}10)$$

and

$$a_y = -(q/m)v_x B_z \qquad (15\text{-}11)$$

respectively. The acceleration is perpendicular to the velocity, a condition indicative of a particle whose direction of travel is changing while the magnitude of its velocity remains constant.

If the magnetic field is uniform, and therefore is the same for all positions of the particle, then the magnetic force is centripetal in nature and the particle moves in a circle. The radius of the circle is $R=mv/(qB_z)$ and the period of revolution is $T=2\pi R/v=2\pi m/(qB_z)$. Here v is the speed of the particle, a constant.

Problem 1. A particle has charge $q=1.6\times 10^{-19}$ C, mass $m=1.7\times 10^{-27}$ kg, and moves in a uniform magnetic field of 1.1 T, in the positive z direction. At time t=0, it is at the origin and has velocity $\vec{v}=6\times 10^5 \hat{i}$ m/s.

a. Use the program of Fig. 7-1 to find the position of the particle every 4×10^{-9} s from t=0 to the time it makes half a revolution. Use $\Delta t=1\times 10^{-10}$ s and plot the orbit.
b. Measure the radius of the orbit from the graph and compare the result with $mv/(qB_z)$.
c. From the data generated in part a, estimate the time the particle takes to make one complete revolution and compare the result with $2\pi m/(qB_z)$.
d. How can the initial conditions be changed so that the position of the center of the circular path is at a different place but the radius is the same?
e. How can the initial conditions be changed so the radius of the circular path is different but the location of the center is the same?

Often, in high energy physics experiments, particles are caused to move in magnetic fields for the express purpose of finding their momenta from measurements of the radii of curvature of the orbits. Magnets are also used to bend beams of particles and so control their trajectories.

There are other situations for which the magnetic force is centripetal in nature and the orbits are circular. These occur when the magnetic field depends only on r, the distance from the z axis to the particle. For a circular orbit of radius R, the acceleration is v^2/R and, since $\vec{v}$ and $\vec{B}$ are perpendicular to each other, Newton's second law yields

$$mv^2/R = qvB(R). \qquad (15-12)$$

This equation can, in principle, be solved for the radius of the orbit, given the speed v. A numerical technique was discussed in section 5.3. If the particle does not start out with v and R related as in Eqn. 15-12, then the orbit is not circular.

<u>Problem 2.</u> Suppose the magnetic field is given by $\vec{B}=50r\hat{k}$ where r is the distance from the z axis. $\vec{B}$ is in tesla and r in meters. A particle has charge $q=1.6\times10^{-19}$ C, mass $m=1.7\times10^{-27}$ kg, and speed $v=6\times10^5$ m/s.

a. Use Eqn. 15-12 to calculate the radius R for which the magnetic field keeps the particle in a circular orbit.

b. Start the particle at x=R, y=0 with velocity $\vec{v}=-v\hat{j}$ and find its subsequent motion. Plot its position every 1×10^{-8} s from t=0 to $t=1.1\times10^{-7}$ s. Use $\Delta t=2\times10^{-10}$ s.

c. Start the particle at x=.5R, y=0 with velocity $\vec{v}=-v\hat{j}$, and find its subsequent motion. Plot its position every 1×10^{-8} s from t=0 to $t=1.1\times10^{-7}$ s. Use $\Delta t=2\times10^{-10}$ s.

Whether the orbit is circular or not, the magnetic field does not change the speed of the particle. The magnetic force is perpendicular to the velocity and it does no work.

<u>Problem 3.</u>
a. For the situation described in part b of problem 2, calculate $v=(v_x^2+v_y^2)^{\frac{1}{2}}$ for the times t=0, 4×10^{-8} s, and 8×10^{-8} s. Is the speed constant to within computational accuracy?

b. For the situation described in part c of problem 2, calculate $v=(v_x^2+v_y^2)^{\frac{1}{2}}$ for the times t=0, 4×10^{-8} s, and 8×10^{-8} s. Is the speed constant to within computational accuracy?

If B is proportional to 1/r where r is the distance from the z axis, then the radius of the orbit no longer appears in Eqn. 15-12. As long as the speed obeys Eqn. 15-12, however, the particle goes into a circular orbit and the radius is equal to its initial distance from the axis. If the speed does not obey Eqn. 15-12, then the orbit is not circular.

<u>Problem 4.</u> The magnetic field, in tesla, is given by

$$\vec{B} = (8.25 \times 10^{-3}/r)\hat{k}$$

where r is the distance, in meters, from the z axis. Consider a particle with charge $q=1.6 \times 10^{-19}$ C and mass $m = 8 \times 10^{-26}$ kg in this field.

a. Use Eqn. 15-12 to calculate the speed v for which the orbit is circular.
b. Start the particle at $x=5 \times 10^{-4}$ m, $y=0$ with velocity $\vec{v}=-v\hat{j}$ and plot its position every 1×10^{-8} s from $t=0$ to $t=1 \times 10^{-7}$ s. Use $\Delta t = 2.5 \times 10^{-10}$ s. Verify that the orbit is circular by computing $(x^2+y^2)^{\frac{1}{2}}$ for times $t=0$, 5×10^{-8} s, and 1×10^{-7} s.
c. Start the particle at $x=3 \times 10^{-3}$ m, $y=0$ with the same velocity and plot its position every 5×10^{-8} s from $t=0$ to $t=6 \times 10^{-7}$ s. Use $\Delta t = 2.5 \times 10^{-9}$ s.
d. Start the particle at $x=3 \times 10^{-3}$ m, $y=0$ with velocity $\vec{v}=-(v/2)\hat{j}$ and plot its position every 6×10^{-8} s from $t=0$ to $t=9 \times 10^{-7}$ s. Use $\Delta t = 3 \times 10^{-9}$ s.

15.4 <u>Motion in Crossed Electric and Magnetic Fields</u>

We take the electric field to be in the x direction and, as before, take the magnetic field to be in the z direction. Then Eqns. 15-2 and 15-3 become

$$a_x = (q/m)(E_x + v_y B_z) \qquad (15\text{-}13)$$

and
$$a_y = -(q/m)v_x B_z \qquad (15\text{-}14)$$

respectively.

If the fields are both uniform, Newton's second law can be integrated and the solution written in analytic form. The motion can be thought of as a combination of two simple motions. First, the particle goes around a circle, just as when there is no electric field. Now, however, the center of the circle moves at a constant velocity.

For an electric field in the x direction, the center of the circle moves in the y direction.

Qualitatively different types of motion occur if the center of the circle moves a distance which is greater or less than a diameter in the time it takes for the particle to complete one revolution. The velocity of the center of the circle can be changed by changing the strength of the electric field.

<u>Problem 1.</u> Let $\vec{B}=1.2\hat{k}$ T and consider a particle with charge $q=1.6\times10^{-19}$ C and mass $m=1.7\times10^{-26}$ kg. It starts at the origin with velocity $\vec{v}=-5\times10^{4}\hat{j}$ m/s. For each of the following electric fields, find the trajectory of the particle. Plot its position every 2×10^{-8} s from t=0 to $t=6\times10^{-7}$ s. Use $\Delta t=1\times10^{-9}$ s.
a. $\vec{E}=1\times10^{4}\hat{j}$ V/m.
b. $\vec{E}=3\times10^{4}\hat{j}$ V/m.
c. $\vec{E}=9\times10^{4}\hat{j}$ V/m.

According to Eqns. 15-13 and 15-14, if $E_x=-v_y B_z$ and the particle is started with its velocity in the y direction, then both a_x and a_y vanish. The velocity is a constant and the particle continues to move straight along a line parallel to the y axis. This condition of zero force is often used to measure the velocities of particles emitted in radioactive decays. The fields are tuned until the particle goes straight through the apparatus. Then v=E/B.

<u>Problem 2.</u> For what electric field does the particle of problem 1 move in a straight line? Is straight line motion, in some sense, intermediate between the motions of part b and part c of problem 1?

15.5 <u>Magnetic</u> <u>Forces</u> <u>and</u> <u>Torques</u> <u>on</u> <u>Circuits</u>

A current carrying loop in a magnetic field experiences a force which is the sum of the forces acting on the moving charges which make up the current. This sum can be written

$$\vec{F} = I \oint d\vec{\ell} \times \vec{B} \qquad (15\text{-}15)$$

where I is the current in the loop, $d\vec{\ell}$ is the vector from one end to the other of an infinitesimal segment of the loop, and $\vec{B}$ is the magnetic field at the integration point. $d\vec{\ell}$ is in the direction of the current and only magnetic fields with sources external to the circuit need be considered. The circuit exerts no net force on itself, although one part may exert a force on another part. We assume the circuit is rigid.

If the only force acting on the circuit is the magnetic force, then this force, divided by the mass of the circuit, gives the acceleration of the center of mass. Given the initial position and velocity, Newton's second law can be integrated to find the subsequent motion of the circuit. We do not, however, pursue that problem here, but rather we concentrate on finding the force on the circuit.

If the field is uniform, the force vanishes. This is easy to understand in terms of Eqn. 15-15. For a uniform field

$$\vec{F} = I \left[\oint d\vec{\ell} \right] \times \vec{B} \qquad (15\text{-}16)$$

and this vanishes because $\oint d\vec{\ell} = 0$. The integral represents a sum of vectors which, when drawn tail to head, necessarily form a closed geometric figure, the loop itself.

If, on the other hand, the field is not uniform, there may be a non-vanishing magnetic force on the circuit.

Problem 1. A square loop of wire carries current of 0.5 A and lies in the x,y plane. Two sides are parallel to the x axis and extend from x=0 to x=0.12 m. The other sides are parallel to the y axis and extend from y=-0.06 m to y=0.06 m. The current flows in the counterclockwise direction when viewed from the positive z axis. For field points in the x,y plane, take the magnetic field to be $\vec{B} = 0.8\hat{j} + 1.5x\hat{k}$ tesla for x in meters.* Find the net force on the loop by considering each side of the square separately as follows.

---

* For the sake of completeness, we note that the field must also have an x component which depends on z if it is to obey $\oint \vec{B} \cdot d\vec{\ell} = 0$ for a closed contour with zero current through it. This component may vanish on the x,y plane and is of no concern to us as long as we need to know the field only in that plane.

a. Along the line y=-0.06 m, $\vec{d\ell}=dx\hat{i}$ and $\vec{d\ell}\times\vec{B}=(0.8\hat{k}-1.5x\hat{j})dx$. Use the program of Fig. 8-1, with N=10, to evaluate the integral. Use the limits of integration to describe the direction of $\vec{d\ell}$: the variable of integration runs from x=0 to x=0.12 m.

b. Along the line x=0.12 m, $\vec{d\ell}=dy\hat{j}$ and $\vec{d\ell}\times\vec{B}=1.5x\hat{i}\,dy$. Use the program of Fig. 8-1 to evaluate the integral. The lower limit is y=-0.06 m and the upper limit is y=0.06 m.

c. Along the line y=0.06 m, $\vec{d\ell}=dx\hat{i}$ and $\vec{d\ell}\times\vec{B}=(0.8\hat{k}-1.5x\hat{j})dx$. Use the program of Fig. 8-1 to evaluate the integral. The lower limit of integration is x=0.12 m and the upper limit is x=0.

d. Along the line x=0, $\vec{d\ell}=dy\hat{j}$ and $\vec{d\ell}\times\vec{B}=0$. The contribution of this side to the total force is zero. Calculate the net force on the loop by vectorially summing the contributions of the four sides.

e. All of the integrals can be evaluated easily in analytic form. Do this and show that $\vec{F}=0.0108\hat{i}$ N.

As another example, we consider a circular loop of radius R, lying in the x,y plane, centered at the origin, and carrying current I. The current flows in the counterclockwise direction when the circuit is viewed from the positive z axis. The integral of Eqn. 15-15 is around the circular loop. We have already described such integrals in connection with finding the magnetic fields of circular current loops and the reader is referred to section 14-2 for details.

We use the angle θ, measured from the positive x axis, to describe the position of a point on the loop. Then $\vec{r}=R\cos\theta\hat{i} + R\sin\theta\hat{j}$ is such a point and $\vec{d\ell} = \vec{dr} = R(-\sin\theta\hat{i} + \cos\theta\hat{j})\,d\theta$. Now $\vec{d\ell}\times\vec{B} = \left[RB_z(\cos\theta\hat{i} + \sin\theta\hat{j}) - R(B_x\cos\theta + B_y\sin\theta)\hat{k}\right]d\theta$ so

$$F_x = IR\int B_z\cos\theta\,d\theta, \qquad (15\text{-}17)$$

$$F_y = IR\int B_z\sin\theta\,d\theta, \qquad (15\text{-}18)$$

and
$$F_z = -IR\int (B_x\cos\theta + B_y\sin\theta)\,d\theta. \qquad (15\text{-}19)$$

These integrals can be evaluated using the program of Fig. 8-1. θ is the variable of integration and it runs from 0 to 2π radians.

Problem 2. Consider a circular loop lying in the x,y plane, centered at the origin, and carrying a current of 0.5 A. The magnetic field is given by $\vec{B}=0.8\hat{j} + 1.5x\hat{k}$ tesla as in problem 1.

a. Find the value of the radius R so that the area enclosed by the circular loop is the same as the area enclosed by the square loop of problem 1.

b. Use the program of Fig. 8-1, run three times, to evaluate F_x, F_y, and F_z. In order to carry out the evaluation, we must write the integrands as functions of θ. In particular, this means making the substitution $x=R\cos\theta$ in the expression for the magnetic field. Use N=10.

If two circuits carry the same current and enclose the same area, the magnetic force on them is nearly the same, regardless of their shapes.

c. Verify this statement for the circuits of problems 1 and 2.

One of the important uses of magnetic fields is to provide a torque on a current loop, thereby tending to turn it. The total torque acting is responsible for a change in the angular momentum of the loop and this phenomenon is exploited to make many types of electrical meters and instruments. We concentrate here on the calculation of the magnetic torque and do not discuss the resulting rotational motion.

To calculate the torque acting on the complete current carrying circuit, we calculate the torque on each infinitesimal segment and add. We choose to calculate the torque about the origin. If $\vec{r}$ is the position of some element $d\vec{\ell}$ of the circuit, then the torque on the circuit is given by

$$\vec{\tau} = I\oint \vec{r}\times\left[d\vec{\ell}\times\vec{B}(\vec{r})\right] \qquad (15\text{-}20)$$

where the integral is around the circuit.

Problem 3. Consider the loop described in problem 1. It carries current of 0.5 A, has two sides parallel to the x axis, extending from x=0 to x=0.12 m and two sides parallel to the y axis, extending from y=-0.06 m to y=0.06 m. The loop lies in the x,y plane and the current flows in a counterclockwise direction when the loop is viewed from the positive z axis. Suppose the magnetic field is given by

$\vec{B} = 0.8\hat{j} + 0.95x\hat{k}$ tesla with x in meters.

a. Show that the components of the torque are given by

$$\tau_x = -0.8 \frac{aI}{2}\left[\int_0^a dx - \int_a^0 dx\right] \quad ,$$

$$\tau_y = -0.8\, I\left[\int_0^a x\, dx + \int_a^0 x\, dx\right] \quad ,$$

and

$$\tau_z = -0.95\, I\left[\int_0^a x^2\, dx + a\int_{-a/2}^{a/2} y\, dy + \int_a^0 x^2\, dx\right] \quad ,$$

respectively. Here a is the length of one of the sides of the square.

b. Use the program of Fig. 8-1 to evaluate these integrals. Obtain three significant figure accuracy. Find the components of the torque.

A current distribution can be characterized by a magnetic dipole moment. For a circuit which lies in a plane, the vector dipole moment $\vec{\mu}$ has a magnitude which is given by the product of the current in the circuit and the area enclosed by the circuit. The direction of $\vec{\mu}$ is perpendicular to the plane of the circuit and is directed such that, if the fingers of the right hand curl around the circuit in the direction of current flow, then the thumb points in the direction of the dipole moment. For a current loop in a uniform field, the torque can be written in terms of the dipole moment of the loop. The result is

$$\vec{\tau} = \vec{\mu} \times \vec{B} \qquad (15\text{-}21)$$

where the field is evaluated for some point within the loop. Eqn. 15-21 is exact for a uniform field and an approximation for a non-uniform field.

Problem 4. Evaluate $\vec{\mu} \times \vec{B}$ for the loop and field of problem 3. Evaluate $\vec{B}$ at the center of the loop. Compare the result for the torque with the answer to part b of problem 3.

A more complicated problem arises from the interaction between the field of a long, straight wire and a current loop. Consider the situation shown in Fig. 14-5. The loop is square with side a and lies in the x,y plane, centered at the origin. We assume it carries current I_1, flowing in the counterclockwise direction when viewed as shown in the figure. The straight wire carries current I_2, in the positive z direction, and pierces the plane of the loop at $x'\hat{i}$.

We calculate the torque, about the origin, exerted by the straight wire on the loop. The magnetic field of the straight wire is given by Eqn. 14-25. This expression, $\vec{r}=x\hat{i}+y\hat{j}$, and $d\vec{\ell}$ are substituted into Eqn. 15-20 and the line integral, around the loop in the direction of I_1, is evaluated.

Across the top $y=a/2$, $d\vec{\ell}=dx\hat{i}$, and the integral runs from $+a/2$ to $-a/2$. Down the left side $x=-a/2$, $d\vec{\ell}=dy\hat{j}$, and the integral runs from $+a/2$ to $-a/2$. Across the bottom $y=-a/2$, $d\vec{\ell}=dx\hat{i}$, and the integral runs from $-a/2$ to $+a/2$. Up the right side $x=a/2$, $d\vec{\ell}=dy\hat{j}$, and the integral runs from $-a/2$ to $+a/2$.

The results are

$$\tau_x = \frac{\mu_0 I_1 I_2}{2\pi} \left[-\frac{a}{2}\int_{-a/2}^{a/2} \frac{(x-x')\,dx}{(x-x')^2+(a/2)^2} + \int_{-a/2}^{a/2} \frac{y^2\,dy}{(\frac{a}{2}+x')^2+y^2} \right. $$
$$\left. -\frac{a}{2}\int_{-a/2}^{a/2} \frac{(x-x')\,dx}{(x-x')^2+(a/2)^2} - \int_{-a/2}^{a/2} \frac{y^2\,dy}{(\frac{a}{2}-x')^2+y^2} \right]$$

and

$$\tau_y = \frac{\mu_0 I_1 I_2}{2\pi} \left[\int_{-a/2}^{a/2} \frac{x(x-x')\,dx}{(x-x')^2+(a/2)^2} - \frac{a}{2}\int_{-a/2}^{a/2} \frac{y\,dy}{(\frac{a}{2}+x')^2+y^2} \right.$$
$$\left. -\int_{-a/2}^{a/2} \frac{x(x-x')\,dx}{(x-x')^2+(a/2)^2} - \frac{a}{2}\int_{-a/2}^{a/2} \frac{y\,dy}{(\frac{a}{2}-x')^2+y^2} \right]. \qquad (15\text{-}22)$$

Here we have, in some instances, inverted the limits of integration and changed the sign in front of the integral. Note that some of the integrals cancel each other and that, for this situation, the top and bottom of the loop do not contribute to the y component of the torque.

Problem 5.   Consider the situation just described.
a. Use Eqns. 14-25 and 15-20 to derive the expressions given by Eqn. 15-22. Omit those integrals which cancel each other.
b. Take I_1=2 A, I_2=1.6 A, a=0.3 m, x'=0.1 m, and use the program of Fig. 8-1, with N=50, to evaluate the integrals in Eqn. 15-24. Calculate the torque on the square loop.
c. The expressions for the components of the torque are also valid if the straight wire pierces the plane of the loop at a point outside the loop. Repeat the calculation of part b with x'=-0.5 m and all other quantities the same.

Chapter 16
TIME DEPENDENT MAGNETIC FIELDS

Faraday's law is discussed and applied to circuits with self-inductance and to circuits coupled by mutual inductance. Problems include calculations of the emf generated when current first starts flowing and the emf of an inductor in a circuit with a sinusoidally varying voltage source. Energy balance for the circuit is considered. These discussions and problems supplement Chapters 35 and 36 of PHYSICS, Chapters 32 and 33 of FUNDAMENTALS OF PHYSICS, and similar material in other texts. The technique given here can easily be extended so that circuits with capacitance can be considered. The problems presented, and others, can be considered in connection with the text book or as part of a laboratory exercise in which various electrical quantities are measured.

16.1 <u>Faraday's Law</u>

Electric and magnetic fields produced by the same set of moving charges are related to each other. The relationship is not a direct one between the fields, but rather, in one form, it is expressed in terms of certain integrals of the fields. We consider one of the relations, known as Faraday's law.

We consider a portion S of a surface as, for example, that shown in Fig. 16-1.

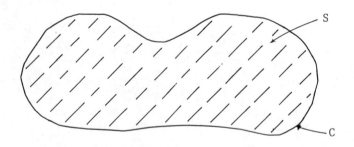

Fig. 16-1 An open surface used to illustrate Faraday's law.

It is not closed. That is, the surface does not surround a volume and the surface itself has a boundary, the line C. The law we are about to describe is given in terms of the surface integral of the normal component of the magnetic field over any surface, such as S, and the line integral of the tangent component of the electric field around the boundary C of S.

226

If there exists a magnetic field at points on the surface, we can calculate the magnetic flux Φ_B through the surface. This flux is defined by

$$\Phi_B = \int B_n \, dS = \int \vec{B} \cdot \vec{dS} \qquad (16-1)$$

Here B_n is the component of the field in a direction perpendicular to the surface at the integration point and dS is an infinitesimal element of surface. $\vec{dS}$ is a vector in a direction normal to the surface. Thus $\vec{B} \cdot \vec{dS} = B_n \, dS$. There is an arbitrariness in the sign of the flux since, at each point on the surface, there are two normal vectors, directed opposite to each other. Either may be chosen as the direction of $\vec{dS}$. Once chosen for one point on the surface, however, the direction of $\vec{dS}$ must be chosen consistently for all other points. For the surface shown, all $\vec{dS}$ vectors must point into the page or all must point out of the page.

We also evaluate the line integral $\oint \vec{E} \cdot \vec{d\ell}$ around the boundary C of the surface S. $\vec{d\ell}$ is an infinitesimal displacement along the boundary and we choose its direction so that it is related to the direction of $\vec{dS}$ by a right hand rule. When the fingers of the right hand curl around the boundary in the direction of $\vec{d\ell}$, then the thumb points in the direction of $\vec{dS}$. The integral is called the electromotive force (or emf) for the contour C. It acts just like the emf of a battery or some other source.

We have now defined the integrals of interest and can write down Faraday's law. It is

$$\oint \vec{E} \cdot \vec{d\ell} = -d\Phi_B/dt. \qquad (16-2)$$

The fields are produced by charges. Whenever the charges move in such a way that there is a magnetic flux through some surface and that flux changes with time, then there is also an emf around the boundary of the surface. In the next section, we consider time changing magnetic fields produced by currents flowing in closed loops.

16.2 Circuits with Inductance

We apply Faraday's law to magnetic fields produced by current in wires. For any open surface we consider, a steady current produces a constant magnetic flux and the emf is zero. It is only when the current changes with time that the flux changes and there is a net emf around the boundary of the surface.

If the boundary of the surface is formed by a conducting wire, the emf which accompanies a changing magnetic flux drives a current through the wire. If there are no other sources of emf present, the current flows in the direction of $d\vec{\ell}$ if $\oint \vec{E} \cdot d\vec{\ell}$ is positive and flows in the opposite direction if that integral is negative.

To start the study of induced emfs in circuits, we consider a square loop of wire as shown below in Fig. 16-2. A magnetic field is created by current I

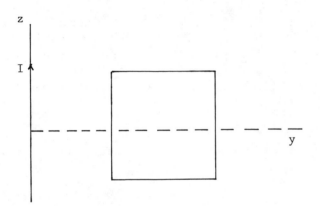

Figure 16-2. A square conducting loop in the magnetic field of a long, straight wire.

in a nearby, long, straight wire and the current is caused to vary with time. As a consequence, the magnetic field varies with time, and so does the magnetic flux through the square loop. There is an emf around the loop and the emf drives a current in the loop.

So far the magnetic field we have been considering is due to the current in the long, straight wire. Current in the loop itself creates a magnetic field and a magnetic flux through the loop, but we neglect these for now. We consider them later.

The magnetic field at every point within the loop is proportional to the current I in the straight wire and therefore, so is the magnetic flux through the loop. We write

$$\Phi_B = MI \qquad (16\text{-}3)$$

where the constant of proportionality M is called the mutual inductance for the system composed of the straight wire and the loop. It can be calculated as the flux through the loop when 1 A flows in the straight wire and its SI unit is called the henry.

In terms of the mutual inductance

$$d\Phi_B/dt = M\, dI/dt \qquad (16\text{-}4)$$

and the emf ε around the loop is given by

$$\varepsilon = -M\, dI/dt. \qquad (16\text{-}5)$$

M is always positive. It is usual, in calculating M, to take the normal to the surface to be roughly in the same direction as the magnetic field so that $\vec{B}\cdot d\vec{S}$ is positive over most of the surface. This defines, by the right hand rule, the direction of positive emf and positive current.

Problem 1. As shown in Fig. 16-2, a long straight wire lies along the z axis and carries current I, which flows in the positive z direction. A square loop lies in the y,z plane and is bounded by the lines z=-0.11 m, z=0.11 m, y=0.1 m, and y=0.32 m. In the figure, the x axis is directed out of the page.

a. Use Ampere's law and the right hand rule to show that the magnetic field at the general point y,z in the plane of the loop is given by

$$\vec{B} = -\frac{\mu_0 I}{2\pi}\frac{1}{y}\hat{i}.$$

b. Take $d\vec{S}$ to be in the negative x direction and show that

$$\Phi_B = \frac{\mu_0 I}{2\pi}\iint \frac{dy\, dz}{y}.$$

Note that Φ_B is proportional to I. The integral can be evaluated by dividing the area of the loop into strips parallel to the z axis. Each strip is 0.22 m in length and has width dy. So

$$\Phi_B = 0.22\,\frac{\mu_0 I}{2\pi}\int_{0.1}^{0.32}\frac{dy}{y}.$$

Use the program of Fig. 8-1, with N=16, to evaluate this integral. Find the value of M for the straight wire and square loop.

c. Suppose the current in the straight wire is given by I=0.55t where I is in amperes and t is in seconds. Find the emf around the loop. What is the direction of the emf?

d. Suppose the resistance of the loop is 125 Ω. What is the current in the loop? In which direction does it flow?

The current in the loop also creates a magnetic field and a magnetic flux through the loop. If this current changes with time, so does the magnetic flux and an additional emf exists around the loop. This emf further influences the current.

Again the emf is proportional to the rate of change of the current and we write

$$\varepsilon = \pm L \, dI/dt \qquad (16\text{-}6)$$

where I is the current in the loop and L is the constant of proportionality, called the self-inductance of the loop. L is always taken to be positive and the sign in Eqn. 16-6 must be chosen so that ε is positive for an emf in the same direction as other positive emfs in the circuit.

Self-inductances are difficult to calculate from first principles. In principle, one should be able to perform the calculation as was done for mutual inductances. That is, assume unit current in the loop, then calculate the magnetic flux through the loop. The difficulty arises since the calculation must take into account the thickness of the wire and the magnetic field inside the wire itself.

We do not attempt a calculation of self-inductance for the square loop, but instead give the result. If a is the length of a side of the square, measured along the center of the wire, and if the wire has circular cross section with radius r, then

$$L = 8 \times 10^{-7} a \left[\ln \frac{a}{r} - 0.52401 \right]. \qquad (16\text{-}7)$$

L is in henries, a and r are in meters.

We now develop the equation which describes changes in the current in the loop when the current in the straight wire changes. The emf which drives the loop current is related to changes in the loop current itself as well as to changes in the current in the straight wire.

Let I_1 be the current in the straight wire, positive for current flowing in the positive z direction, and let I_2 be the current in the loop, positive for current flowing in the counterclockwise direction when the loop is viewed as it is in Fig. 16-2. Then the emf around the loop is

$$\varepsilon = M\, dI_1/dt - L\, dI_2/dt \qquad (16\text{-}8)$$

where ε is taken to be positive for an emf in the counterclockwise direction when the loop is viewed as it is in Fig. 16-2. You should verify the sign in Eqn. 16-8 by using the right hand rule. Assume that dI_1/dt is positive and show that the resulting emf is positive, then assume that dI_2/dt is also positive and show that the resulting contribution to the emf is negative.

If the resistance of the loop is R and there are no other sources of emf acting, then $\varepsilon = I_2 R$ and substitution into Eqn. 16-8 yields

$$I_2 R = M\, dI_1/dt - L\, dI_2/dt. \qquad (16\text{-}9)$$

In order to apply this equation, we think of I_1 as being a given function of time and I_2 as being an unknown function of time, to be found. Eqn. 16-9 may be rearranged to read

$$dI_2/dt = (M/L)\, dI_1/dt - (R/L)\, I_2. \qquad (16\text{-}10)$$

This equation has the same form as Newton's second law for a particle moving in one dimension and subjected to both a time dependent and a velocity dependent force. I_2 corresponds to the velocity and the left side of Eqn. 16-10 corresponds to the acceleration of the particle. The right side of the equation corresponds to the force divided by the mass of the particle. The first term is like a time dependent force and the second term is like a velocity dependent, drag force.

The program of Fig. 6-1 can be used to integrate Eqn. 16-10. I_2 is stored in X2 and the function evaluated at line 150 is

$$(M/L)\, dI_1/dt - (R/L)*X2 \rightarrow X6$$

Here the appropriate numerical values must be substituted for (M/L) and (R/L) and dI_1/dt must be given as a function of time. The time is stored in X3 while X4 contains the interval width Δt for the integration and X5 contains the number of intervals considered by the machine before the result is displayed.

X1 is not needed and all reference to it can be eliminated (at lines 100 and 200). Line 170 can be eliminated in its entirety. This improves the running time of the program somewhat. If X1 is not eliminated, an initial value must be entered. An initial value of 0 gives a nice physical meaning to the number stored in X1. If X1 is initially zeroed, then at a later time it contains $\int I_2\, dt$, which is the net amount of charge that has passed any point on the loop since the time t=0. It is sometimes of interest to compute this quantity.

If I_1 varies linearly with time, dI_1/dt is constant and a possible solution to Eqn. 16-10 is

$$I_2 = (M/R)\, dI_1/dt, \qquad (16-11)$$

a constant. This is, in fact, the so-called steady state solution. It does not depend on the self-inductance L of the loop, but only on the mutual inductance M. The value of I_2 given by Eqn. 16-11 may not be the initial value of I_2, in which case I_2 changes with time until the steady state value is reached. The situation is analogous to the one dimensional motion of a body subjected to a drag force proportional to its speed. The steady state value of I_2 corresponds to the terminal velocity of the body.

Problem 2. Consider the straight wire and loop of problem 1. Take the radius of the wire to be 0.04 cm and the resistance of the loop to be 125 ohms. Suppose I_1=0.55t A, with t in seconds.

a. Use Eqn. 16-7 to find the self-inductance of the loop.
b. Use Eqn. 16-11 to find the value of the steady state current I_2. Take M to have the value you found in problem 1.
c. Take the initial value of I_2 to be 0 and plot I_2 every 5×10^{-9} s from t=0 to t=6×10^{-8} s. Use the program of Fig. 6-1 with $\Delta t = 1\times 10^{-10}$ s.
d. Take the initial value of I_2 to be twice the steady state value and plot I_2 every 5×10^{-9} s from t=0 to t=6×10^{-8} s. Use the program of Fig. 6-1 with $\Delta t = 1\times 10^{-10}$ s.
e. For each of the situations of parts c and d, estimate from the graphs the time

taken for the current I_2 to go 80% of the way from its initial value to the steady state value.

Although the final steady state value of I_2 does not depend on L, the time the current takes to reach the steady state value does. The self-inductance tends to inhibit changes in the current I_2 and the larger L, the longer the system takes to reach steady state.

Problem 3.
a. Rework part c of problem 2 for a loop with double the self-inductance found in part a of problem 2. Take M and dI_1/dt to be the same. L can be changed without changing M if the radius of the wire is changed.
b. Rework part c of problem 2 for a loop with half the self-inductance found in part a of that problem.
c. For parts a and b, use the graphs to estimate the time for the current to reach 80% of its steady state value. Verify that the results substantiate the statement made above about the dependence of this time on L.

Inductance is an important part of many electrical circuits in which the current changes. The inductance may arise because the circuit bounds an area through which there is a changing magnetic flux or because a circuit element with large inductance, such as a coil of wire, is purposely included in the circuit by the designer.

We consider the simplest circuit containing a voltage source, a resistor, and an inductor. They are placed in series as shown in Fig. 16-3.

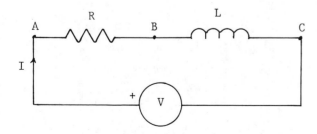

Fig. 16-3. A single loop circuit containing voltage source V, resistance R, and inductance L, in series.

We use the following sign convention. V is positive if the terminal of the source marked with a + is at a higher potential than the other terminal. Otherwise, V is negative. The current I is positive if it flows in the direction shown by the arrow.

L is meant to account for the total inductance of the circuit and R is meant to account for the total resistance.

Kirchoff's loop rule holds and, if we arbitrarily assign the value 0 to the potential at the negative terminal of the source, then the potential at A is V, the potential at B is V-IR, and the potential at C is V-IR-L dI/dt. This last expression must vanish since the potential at C is the same as the potential at the negative terminal of the source. Thus, the current in the loop obeys the equation

$$dI/dt = -(R/L)I + (V/L). \qquad (16\text{-}12)$$

Note that if the current is positive (flows in a clockwise direction in Fig. 16-3) and increasing in magnitude, then the inductor acts as a source of emf directed against the current and point B is at a higher potential than point C. On the other hand, if the current is positive and decreasing in magnitude, then the inductor acts as a source of emf directed with the current and point C is at a higher potential than point B. In general, the presence of the inductor tends to inhibit changes in the current.

We investigate the circuit for two different situations. First, we consider a source of constant voltage, which is turned on at t=0. Eventually, the current reaches its steady state value V/R. This can be found from Eqn. 16-12 by placing dI/dt=0. The current does not immediately rise to its steady state value, however, but rather approaches that value over a time interval which depends on the value of L.

Secondly, we consider a sinusoidally varying source. Again the current is not given by V/R but differs from this value because the inductor acts as another source of emf. Not only is the magnitude of the current different from V/R but the current is out of phase with the voltage source. That is, the current and the voltage across the source reach maximum values, for example, at different times. In this case the voltage source is time dependent and there is no steady state solution to Eqn. 16-12.

The program of Fig. 6-1 can be used to find solutions to Eqn. 16-12. Again I is stored in X2 and the time in X3. X1 is not needed. It may be eliminated from the program or, if retained, it may be started with the value 0. The function evaluated at line 150 is

$$-(R/L)*X2 + (V/L) \rightarrow X6$$

where a numerical value must be substituted for (R/L) and (V/L) must be given as a function of time.

Problem 4. The circuit depicted in Fig. 16-3 has a resistance of 100 Ω. At time t=0, the current is 0 and the voltage source is turned on. At that time the voltage jumps from 0 to 15 V. Use the program of Fig. 6-1 to plot, on the same paper, the current I as a function of time for each of the given values of inductance. Plot I every 4×10^{-4} s from t=0 to t=4.8×10^{-3} s. When I is nearly constant, of course, the graph can be plotted at larger intervals.
a. L=5×10^{-2} h.
b. L=1×10^{-1} h.
c. L=2×10^{-1} h.
d. On the graph, draw the line at I=V/R.
e. On the graph, draw the line which represents a current which is 80% of the steady state value and, for each value of L, estimate the time T required for the current to reach 80% of the steady state value. Plot T as a function of L. According to theory, T should be proportional to L. Does your graph support that relationship? There is nothing special about the 80% line. The same statement can be made about the time to reach any other fraction of the steady state value.

Problem 5. For the circuit of Fig. 16-3, take R=100 Ω and V=$15\sin(1 \times 10^4 t)$ where V is in volts for t in seconds and the argument of the sine function is in radians. Take I=0 at t=0 and, on separate graphs, plot the following quantities every 5×10^{-5} s from t=0 to t=1.5×10^{-3} s. Use t=$2 \, 10^{-6}$ s as the integration interval for parts b and c.
a. Plot V(t)=$15\sin(1 \times 10^4 t)$.
b. Take L=1×10^{-2} h and plot I(t).
c. Take L=3×10^{-2} h and plot I(t).

Notice that the current eventually becomes sinusoidal and that the characteristic time for it to do so is longer for the larger value of L. This is in agreement with the findings of problem 4. The current, however, never becomes V/R and, in fact, the maximum value of I is less for larger L. The current is always changing with time, there is always a back emf across the coil, and I is always less than V/R.

Notice also that the current reaches its maximum value at a time which is later than the time when the voltage across the source reaches its maximum and that the lag is greater for greater L.

Results of the last problem can be used to demonstrate energy balance for circuits with inductance. At any instant of time, the voltage source supplies energy at a rate given by P_s=VI where V is the voltage drop across the source and I is the current through it. Both V and I are functions of time and P_s may change with time. The source supplies energy to the circuit when P_s is positive (V and I in the same direction) and takes energy from the circuit when P_s is negative (V and I in opposite directions).

The rate at which energy is dissipated in the resistor is given by $P_R=I^2R$. This is always a positive number. The resistor always draws energy from the circuit. Energy withdrawal occurs by means of collisions between electrons in the current and the atoms which make up the resistor and the energy transfer results in an increase in the temperature of the resistor.

There is energy stored in the magnetic field created by the current and the amount of energy (compared to the energy stored when I=0) is $\frac{1}{2}LI^2$. The rate at which this energy changes is $d(\frac{1}{2}LI^2)/dt = LI\, dI/dt$.

Energy supplied by the voltage source is either dissipated by the resistor or stored in the magnetic field, so it must be that

$$VI = RI^2 + LI\, dI/dt. \tag{16-13}$$

This equation can be derived mathematically from Eqn. 16-12. Simply multiply that equation by LI and rearrange the terms.

In the next problem, we calculate the individual terms in Eqn. 16-13 for the situation described in problem 5.

Problem 6. Use the results of parts a and b of problem 5 and, on separate graphs, plot the following quantities every 1×10^{-4} s from t=0 to t=1.5×10^{-3} s.

a. VI. Identify the time intervals during which the source is supplying energy to the current and the time intervals during which it is taking energy from the current. Notice that the source may supply energy with V either positive or negative.

b. I^2R. The power dissipated by the resistor is always positive. The resistor never supplies energy to the current.

c. LI dI/dt. This quantity may be calculated from LI dI/dt = VI - I^2R. This is positive when both I and dI/dt have the same sign. Then the magnetic field is increasing in magnitude and energy is being taken from the current to be stored in the field. When I and dI/dt have opposite signs, the field is collapsing and the energy is being returned to the current.

Circuits may be coupled by means of their mutual inductance. Sometimes this happens because two circuits are near each other and the magnetic field, created by the current in one, passes through the area bounded by the other. Problem 1 is an example.

Sometimes a large mutual inductance is purposely created in order to produce a current in one circuit when the current in another circuit changes. This can be done, for example, by means of a transformer. Two coils, one in each circuit, are isolated electrically from each other, but are wrapped on the same iron core. The core insures a large mutual inductance since it concentrates the field lines so that most of them pass through both coils.

We consider the situation depicted in Fig. 16-4. Circuit 1, on the left, has voltage source V, self-inductance L_1, and resistance R_1 while circuit 2, on the right, has self-inductance L_2 and resistance R_2. The circuits are coupled by a mutual inductance M. V is considered to be positive if the terminal marked + is at a higher potential than the other terminal and the currents are considered to be positive if they flow in the directions of the arrows. These are also the directions of positive emfs in the two circuits.

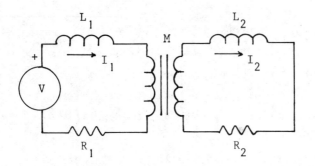

Fig. 16-4. Two circuits coupled by a mutual inductance M.

We assume, for this example, that an increase in I_1 produces an emf in circuit 2 which is in the clockwise direction when viewed as in Fig. 16-4. The emf is positive. An increase in I_2 then produces a positive emf in circuit 1. It is possible to wind the transformer so that the sign of the emf is negative in one circuit for an increasing current in the other circuit. The self-inductance, of course always inhibits changes in the current of its own circuit.

For circuit 1,

$$V - L_1 \, dI_1/dt + M \, dI_2/dt - I_1 R_1 = 0 \qquad (16\text{-}14)$$

and, for circuit 2,

$$-L_2 \, dI_2/dt + M \, dI_1/dt - I_2 R_2 = 0. \qquad (16\text{-}15)$$

These are two simultaneous equations to be solved for the two functions of time, $I_1(t)$ and $I_2(t)$. In preparation for solving them numerically, we manipulate them so that either dI_1/dt or dI_2/dt appears in each equation, but not both. Multiply Eqn. 16-14 by L_2 and Eqn. 16-15 by M, then add the results to obtain

$$L_2 V - L_1 L_2 \, dI_1/dt - L_2 R_1 I_1 + M^2 \, dI_1/dt - M R_2 I_2 = 0. \qquad (16\text{-}16)$$

Now multiply Eqn. 16-14 by M and Eqn. 16-15 by L_1 and add the results to obtain

$$MV + M^2 \, dI_2/dt - M R_1 I_1 - L_1 L_2 \, dI_2/dt - L_1 R_2 I_2 = 0. \qquad (16\text{-}17)$$

These can be rearranged to produce

$$dI_1/dt = -(L_2R_1I_1 + MR_2I_2 - L_2V) / (L_1L_2 - M^2) \qquad (16-18)$$

and
$$dI_2/dt = -(MR_1I_1 + L_1R_2I_2 - MV) / (L_1L_2 - M^2) \qquad (16-19)$$

respectively.

These equations have the same form as Newton's second law in two dimensions, with I_1 and I_2 replacing the two components of the velocity. The program of Fig. 7-1 can be used without modification. I_1 is stored in X3, I_2 is stored in X4 while X5 contains the time t, X6 contains the interval Δt used for the integration, and X7 contains the number of intervals considered by the machine before results are displayed. X1 contains the charge which has passed any point in circuit 1 and X2 contains the charge which has passed any point in circuit 2. For most problems, they are not needed and any reference to them may be removed from the program. If they are retained, start them with values of 0.

At line 150, the right side of Eqn. 16-18 should be used and the line should read

$$-((L_2R_1)*X3+(MR_2)*X4-(L_2V))/(L_1L_2-M^2) \rightarrow X8$$

Similarly, the right side of Eqn. 16-19 is used at line 160 and this line should read

$$-((MR_1)*X3+(L_1R_2)*X4-(MV))/(L_1L_2-M^2) \rightarrow X8$$

In each case, numerical values should be substituted for L_1, L_2, M, R_1, and R_2. V should be replaced by a function of time.

In the following problem, we investigate the current generated in circuit 2 when the voltage source in circuit 1 in turned on. The response of circuit 2 clearly depends on the value of M since the primary emf which drives I_2 is the emf which accompanies changes in I_1.

Problem 7. For the circuit depicted in Fig. 16-4, take L_1=35 mh, L_2=65 mh, R_1=150 Ω, R_2=250 Ω, and suppose that, at time t=0, the voltage source is turned on. At that time I_1=0, I_2=0, and V jumps from 0 to 15 V. For each of the following two values of M, plot I_1 and I_2 as functions of time from t=0 to t=1.5×10^{-3} s. At first, plot points every 5×10^{-5} s, then, when the currents are changing less rapidly, every 1×10^{-4} s. Use the program of Fig. 7-1 with Δt=5×10^{-6} s.

a. M=0.2 mh (weak coupling between the circuits).
b. M=20 mh (stronger coupling between the circuits).

The graphs clearly show I_1 reaching its steady state value of V/R_1=0.1A. At first, when I_1 is changing rapidly, a large I_2 is generated. As I_1 reaches its steady state value, dI_1/dt becomes small and I_2 also decreases. The influence of the self-inductance of circuit 2 is shown clearly by the plots of I_2. At long times, the self-inductance acts to reinforce I_2 as the driving emf dies out.

APPENDICES

The purpose of these appendices is to give help in making the translation from flow charts to specific machine instructions. Machine instructions described in these appendices do not necessarily produce the fastest running programs or the fewest programming steps. They are, however, adequate to produce working versions of the programs described in the main text and, we believe, they are relatively easy for a novice to use. As you become more familiar with your machine and its capabilities, you should try to generate more efficient programs.

The chief parts of a program deal with entering and storing data, recalling data from memory, performing arithmetic and other operations, transferring control from one part of the program to another, and displaying results. In the following appendices, specific machine instructions are given for several widely used machines so that those machines can be programmed to carry out the tasks listed above. The instructions are keyed to flow chart statements to facilitate the translation from flow chart to instruction.

Programming languages for large machines allow the user to name the storage locations. When possible, it is good programming practice to use names to remind you what physical quantity is represented by the number in storage. Thus the x component of the electric field might be stored in EX and the y component of the velocity might be stored in VY. This practice greatly helps in the reading and understanding of programs.

Some hand held calculators allow an approximation to this procedure, at least for input data. A number is stored by an instruction which is labelled by a single letter and the user keeps a record of the correspondence between the letter and the physical significance of the number.

Whatever method you use, you should acquire the habit of noting the physical significance of each storage location used to accept input data. When running the programs of this book on a hand held calculator, it is a good idea to program the machine to display the memory location just prior to stopping to accept data. In this way, you are reminded which number is to be entered.

The programs in this book are written for small calculators. When the calculation is finished, the user stops the machine or does not restart it after copying the last output desired. For most large machines, the user cannot do this and instructions must be written to tell the machine when the calculation is finished. For example, suppose the position of a particle is to be found for a series of ten values of the time. Results are printed and the machine does not stop for each result. The machine must then keep track of the value of the time and stop after the calculation for the final value of the time. This is easily done by means of a conditional transfer statement. When the value of the time is larger than some predetermined number, the machine is instructed to transfer to a STOP statement.

Sample Program

A sample program is given in the second figure of each appendix. It has been included to allow you to see a complete program, written for your machine. As you read the following description of the program, refer to the appropriate flow chart in any one of the appendices.

The program calculates the sum

$$S = \sum_i \cos(x_i)$$

where $x_1=0$, $x_2=\Delta x$, $x_3=2\Delta x$, ... up to some largest value, just less than x_{max}. Δx is stored in X1, x_{max} is stored in X3, and x_i is stored in X2. The sum is collected in X4. X4 is zeroed before the loop is entered and, as each term is evaluated, at line 140, it is added to the previously computed terms. Each time around the loop, X2 is incremented by X1. When X2 becomes equal to or larger than X3, all terms have been included and the result is displayed.

Appendix A
TEXAS INSTRUMENTS MACHINES

The description we give here is for the TI 58 and TI 59 machines. With slight modification, it is valid for other Texas Instruments calculators as well.

Nearly all of the keys of these machines have two uses. A given key has one use when used alone and another when it is used following the key marked 2nd. The first use is marked on the key itself while the second use is marked above the key. In listing keys, we specify the first use by writing the symbol on the key and we specify the second use by writing 2nd followed by the symbol written above the key. For example, we write x^2 and 2nd sin for the two uses of the third key in the third row of the keyboard.

To program these machines, press Lrn, then the keys corresponding to instructions to be executed in order. When programming is finished, press Lrn again to exit the programming mode. To run the program, press CLR followed by RST and R/S. CLR clears pending operations, if any; RST tells the machine to go to line 000; and R/S tells it to start executing instructions. R/S starts the machine when it is stopped and stops it when it is running.

Fig. A-1 gives a summary of machine instructions corresponding to flow chart statements used in this book.

Look first at the ENTER statement. The programmed instruction R/S stops the machine to allow the first number X1 to be keyed in from the keyboard. The operator must then push R/S to start the machine. Immediately the number is stored in memory 01. The machine then encounters a programmed R/S and stops to allow X2 to be keyed in. The operator then presses R/S and X2 is stored in memory 02.

Memory locations are two digit numbers, but if the location number is less than 10 and it is followed by a non-numeric symbol (R/S, for example), only the single digit need be given. To avoid confusion, we always give two digit location numbers.

When the machine encounters STO 01, it stores in memory location 01 whatever number is in the display. For the example given in Fig. A-1, the number X1 is in the

243

Figure A-1. Examples of instructions for Texas Instruments machines.

a. <u>Enter and store</u>

```
     _____
    |                \
    |     ENTER       |
    |     X1, X2      |
    |_____|
```

    R/S   machine stops, user keys in X1, presses R/S to start machine.
    STO 01
    R/S   machine stops, user keys in X2, presses R/S to start machine.
    STO 02

b. <u>Operations</u>

```
    |  5*X1→X2  |
```
5 × RCL 01 = STO 02

```
    |  X1+X2→X3  |
```
RCL 01 + RCL 02 = STO 03

```
    |  (X1+X2)*X3→X4  |
```
RCL 01 + RCL 02 = × RCL 03 = STO 04
              or
(RCL 01 + RCL 02) × RCL 03 = STO 04

```
    |  SIN(X1)→X2  |
```
RCL 01 2nd sin STO 02

```
    |  X1*SIN(X2*X3)
           →X4      |
```
RCL 02 × RCL 03 = 2nd sin × RCL 01 = STO 04
                or
(RCL 02 × RCL 03) 2nd sin × RCL 01 = STO 04

Figure A-1. Cont'd. Texas Instruments.

> INVTAN(X1)→X2

RCL 01 INV 2nd tan STO 02
(Choose degree or radian mode for angle)

> X1↑2→X2

RCL 01 x^2 STO 02

> SQRT(X1↑2
> +X2↑2)→X3

RCL 01 x^2 + RCL 02 x^2 = √x STO 03

> X1↑1.5→X2

RCL 01 y^x 1.5 = STO 02

> f(X1,X2)→X7

5 × RCL 01 + RCL 02 = STO 07

where f(X1,X2)
=5*X1+X2

c. <u>Display</u> <u>or</u> <u>print</u> <u>results</u>

> DISPLAY
> X1, X2

RCL 01 R/S   machine stops, user copies X1, presses R/S to start machine.

RCL 02 R/S   machine stops, user copies X2, presses R/S to start machine.

> PRINT
> X1, X2

RCL 01 2nd Prt
RCL 02 2nd Prt

Figure A-1. Cont'd.   Texas Instruments.

d. <u>Transfer statements</u>

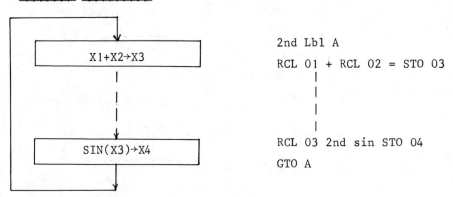

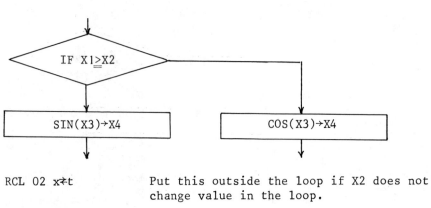

Figure A-1. Cont'd.   Texas Instruments.

e. <u>Additional instructions</u>

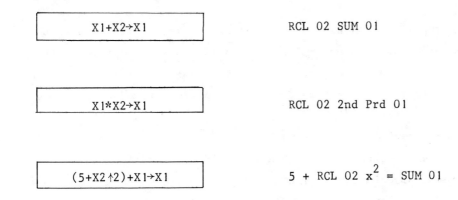

$X1+X2 \rightarrow X1$	RCL 02 SUM 01
$X1*X2 \rightarrow X1$	RCL 02 2nd Prd 01
$(5+X2\uparrow 2)+X1 \rightarrow X1$	$5 +$ RCL 02 $x^2 =$ SUM 01

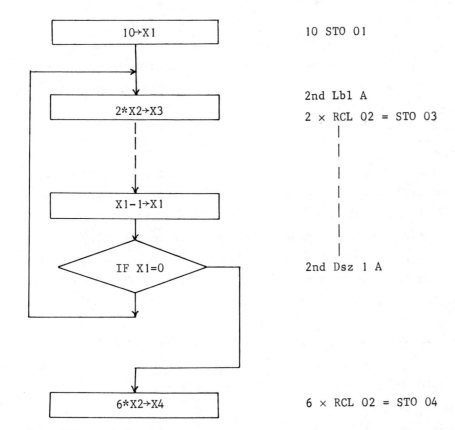

$10 \rightarrow X1$	10 STO 01
$2*X2 \rightarrow X3$	2nd Lbl A $2 \times$ RCL 02 = STO 03
$X1-1 \rightarrow X1$	
IF $X1=0$	2nd Dsz 1 A
$6*X2 \rightarrow X4$	$6 \times$ RCL 02 = STO 04

display because it was keyed in by the operator at the appropriate time. The number also remains in the display where it can be used immediately in a numerical operation. If the storage location previously contained a number, it is erased when the new number is stored.

Numbers are entered into the display by pressing the keys corresponding to their digits. Scientific notation is used by pressing EE followed by one or two digits which give the power of ten. Negative numbers are entered by pressing +/- after the magnitude of the number has been entered and negative powers of ten are entered by pressing +/- just before the magnitude of the exponent.

The next group of statements deals with arithmetic and function operations. All involve at least one recall from memory. The instruction RCL 01 places the number which is currently in memory 01 into the display. The number also remains in memory.

In general, arithmetic operations have the form: put the first number into the display, press the operation key, put the second number into the display, press =. Possible operations are +, -, ×, ÷, and y^x. The first four are the usual arithmetic operations. The last raises the first number to the power given by the second. The first number must be positive. The operation key prepares the machine to accept the second number and tells it what operation to perform, but the operation itself is not performed until the = is encountered. The result of the operation is automatically placed in the display, where it might act as the first number in a new operation.

Numbers are placed in the display from the keyboard, from memory (through use of the RCL instruction), or as a result of an arithmetic operation. The number in the display is called x for purposes of labelling keys. The 5 in the first example of Fig. A-1, part b, is programmed into the display. When the machine encounters the programmed 5, it acts just as if the operator had pressed the 5 key, thereby entering 5 into the display. Other numbers, in this and other examples, are entered into the display be recalling them from memory.

Arithmetic operations can be strung together by means of parentheses.

$$((RCL\ 01 + RCL\ 02) \times RCL\ 03 + RCL\ 04) \div RCL\ 05 =$$

is a legitimate sequence of keys. Within each set of parentheses multiplications and divisions are performed before additions and subtractions. The value in the inner set of parentheses is multiplied by X3 and the result is added to X4. Inner most sets of parentheses are evaluated first, then outer sets.

The right parenthesis plays the role of the = instruction. When it is encountered, pending operations within that set of parentheses are completed. Likewise, a second arithmetic operation may cause pending operations to be completed when there is no doubt about the interpretation of the pending operations. For example, the first addition in RCL 01 + RCL 02 + RCL 03 is completed when the second one is encountered. If there is any doubt about what a sequence of instructions does, check manually using numbers entered from the keyboard, then program the instructions.

Texas Instruments machines are capable of evaluating a number of functions such as x^2, $\sqrt{x}$, $1/x$, $|x|$, the trigonometric functions, and the inverse trigonometric functions. When any of these operators are encountered, they immediately act on the number in the display and so they must be encountered by the machine after the number to be operated on has been placed in the display. The appropriate order of instructions to find the square root of X1, for example, is RCL 01 $\sqrt{x}$. First put the number in display, then find its square root. The fourth example of operations given in Fig. A-1 shows how the sine of an angle is found.

Any function, once evaluated, is in the display and can be used as part of further arithmetic operations, as shown by other examples in Fig. A-1. Notice that the 2nd key must be pressed before the trigonometric keys. For inverse trigonometric functions, the sequence of keys is INV 2nd sin, for example. INV $\ln x$ produces e^x and INV 2nd log produces 10^x where x is the number in the display.

Sequences of keys should not be read as if they were algebraic statements. For example, RCL 01 − RCL 02 2nd sin = × RCL 03 = STO 01 makes no sense as an algebraic statement but it means, as a series of instructions, to put the number in memory 01 into display, subtract from it the sine of the number in memory 02, put the result in the display, multiply the number in the display by the number in memory 03, put the result in the display, and store it in memory 01. Notice that the sine operator acts immediately on the number in the display, even though that number is the second partner in a pending arithmetic operation. The arithmetic operation is not completed until the = instruction is encountered.

Many programs in this book are written with an undefined function in one or more statements. The function is supplied by the user at the time the program is run. For these programs, the user must substitute appropriate instructions in the form of arithmetic and function operators acting on numbers already stored in memory. If the same function is used more than once in a program, it is efficient to evaluate it in a subroutine. See the instruction manual for details.

Instructions appropriate for flow chart display statements recall the number from memory, automatically to the display, then stop the machine with R/S. After the number has been copied, the operator pushes R/S to restart the machine. Printing capability avoids not only the necessity of copying the number by hand but also the necessity of stopping the machine. The instruction 2nd Prt causes the machine to print the number in the display.

In order for transfer instructions to tell the machine what instruction to execute next, instructions can be labelled. We use the 5 letters A, B, C, D, E and the 5 additional letters A´, B´, C´, D´, E´, available by means of the 2nd key, to label instructions. Actually, almost any of the keys can be used to label instructions.

An instruction is labelled by programming the sequence 2nd Lbl A, for example, just before the instruction. The primes are brought into use by the sequence 2nd Lbl 2nd A´. Somewhere else in the program the sequence of instructions might be GTO A. When this is encountered, the machine goes to the instruction labelled A and begins executing instructions there.

Conditional transfers are a little more complicated. The number in display (x) is compared to the number in a special register, called t. Numbers are placed in t by first placing them in the display, then using the x⇄t instruction, which interchanges the contents of the display and t registers. Another number is then placed in the display for purposes of comparison. The comparison itself takes place by means of one of the following four instructions:

$$\text{2nd } x=t$$
$$\text{INV 2nd } x=t \quad (\text{equivalent to } x \neq t)$$
$$\text{2nd } x \geq t$$
or $\quad\text{INV 2nd } x \geq t \quad (\text{equivalent to } x < t)$

followed by a statement label (A, B, etc.). If the comparison statement is true, then the machine goes next to the statement with the stated label. If the comparison statement is false, no transfer is made. The machine ignores the label and goes to the next instruction in sequence.

For the example shown in Fig. A-1, if X1 is in fact greater than or equal to X2, then the machine goes to the instruction labelled A and computes the cosine of X3. If, on the other hand, X1 is less than X2, the machine goes to the next instruction and computes the sine of X3.

If the comparison statement is in a loop and one of the quantities, X1 or X2, does not change value in the loop, that quantity should be placed in t before the machine enters the loop. This saves some running time.

There are five special programming instructions which are available with Texas Instruments machines. Strictly speaking, none are necessary since the same results can be produced by means of other instructions, already given. They save so much running time, however, that we choose to describe them here and give examples in part e of Fig. A-1.

Some arithmetic can be done in memory without recalling both partners in the operation to display. The instruction SUM 01 adds the display number to the content of memory 1 and stores the result in memory 1. The instruction 2nd Prd 01 multiplies the content of memory 1 by the display number and stores the result in memory 1. Subtraction and division are accomplished by means of the instructions INV SUM 01 and INV 2nd Prd 01, respectively. Any other 2 digit memory location number may be used in place of 01 in these instructions.

The fifth instruction has the form 2nd Dsz N A where N is a one digit memory location number from 0 through 9 and A is the label of a program instruction. An integer is first stored in location N before the 2nd Dsz N A instruction is encountered for the first time. Then, every time the machine encounters this instruction, the integer in N is reduced by 1. Following the reduction, the integer is checked to see if it is zero. If it is not zero, the machine next goes to the statement labelled A. If, on the other hand, the integer is zero, the machine goes to the instruction following the 2nd Dsz N A instruction.

The statement is useful when one wants the machine to execute a loop a specified number of times. In the example shown in Fig. A-1, X1 is an integer and the loop, represented by the dotted line, is traversed exactly X1 times. Note that the number of traversals in placed in X1 before the loop is entered and the 2nd Dsz instruction is the last instruction of the loop. The first instruction of the loop is labelled with the label mentioned in the 2nd Dsz instruction.

Fig. A-2 shows a flow chart for the sample program described in the introduction to the appendices. Instructions for Texas Instruments machines are given along side the flow chart. Compare carefully. At line 130 it is not necessary to recall X2 since it happens to be in display then. For other programs, the IF statement might require the recall of a number from memory.

For several of the programs in this book, memory locations need to be computed. For example, we may wish to store numbers in X(3*X2+1), X(3*X2+2), and X(3*X2+3) where X2 is a variable and takes on the values 1, 2, 3, ... in a loop. The numbers in the parentheses are the memory location numbers.

One way to accomplish the storage is as follows. Three storage spaces, say 28, 29, and 30, are set aside to hold the numbers 3*X2+1, 3*X2+2, and 3*X2+3, respectively. The following sequence stores the numbers:

$$\text{RCL } 02 \times 3 = + 1 = \text{STO } 28$$
$$+ 1 = \text{STO } 29$$
$$+ 1 = \text{STO } 30$$

After the first number is put into memory, it also remains in the display and can be added to 1 to produce the next number to be stored. Remember that memory 28 holds the location number of the memory where X(3*X2+1) is to be stored. It does not hold X(3*X2+1) itself. Similarly for the other two locations.

Now the first number, to be stored in location 3*X2+1, is keyed into display. It is placed in the proper memory by means of a process called indirect addressing. The key sequence is STO 2nd Ind 28. The displayed number is stored, not in memory 28, but in the memory whose location number is the number stored in location 28. Similarly the other two numbers are stored by using STO 2nd Ind 29 and STO 2nd Ind 30, respectively. The numbers are recalled to the display by the instructions RCL 2nd Ind 28, RCL 2nd Ind 29, and RCL 2nd Ind 30, respectively.

Figure A-2. Sample program.

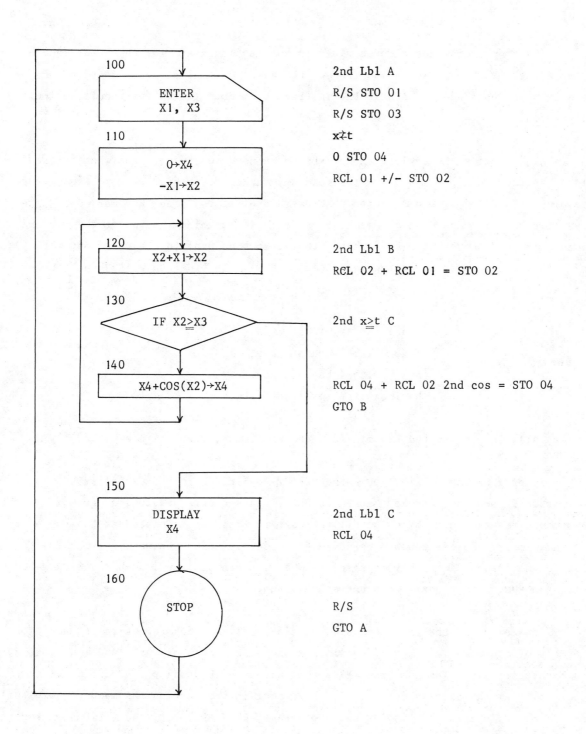

Appendix B
HEWLETT-PACKARD HAND HELD MACHINES

The programming key strokes given in this appendix are for the HP-67 machine. Other Hewlett-Packard calculators (except the 41) are similar, but differ in some details. The user should consult the instruction manual for the particular calculator being used.

Many keys on the Hewlett-Packard keyboard have four uses: one when pressed alone, another when pressed after the gold key marked f, a third when pressed after the blue key marked g, and a fourth when pressed after the black key marked h. The uses of a key are marked on the machine as follows: in black on top of the key (used alone), in gold beneath the key (used with f), in blue beneath the key (used with g), and in black on the front of the key (used with h). In order to write down the key labels, we write

                label on top of key,
                f gold label beneath key,
                g blue label beneath key,
or   h black label on front of key.

For example, labels for the key marked 5 are written 5, f cos, g $\cos^{-1}$, and h y^x.

To program the machine, push the W/PRGM-RUN switch to W/PRGM, press the gold key marked f, and then the key marked CL PRGM. This clears previous programs from the memory and resets the machine so that the next instruction is placed at the top of the program memory. Keys are now pressed in the order in which you want the machine to execute the instructions they represent. When you are finished, push the W/PRGM-RUN switch to RUN. The program is then ready to run.

The memory of the machine has 4 locations called stack registers, in addition to a number of addressable storage registers. The stack registers are labelled X, Y, Z, and T respectively. The number in X is the number which is displayed and we often refer to it as the display number. It is helpful in programming the machine to visualize 4 numbers, one in the display and 3 others stacked above it. As various instructions are carried out, the numbers in the stack registers change and it is worthwhile to visualize the change. For each instruction, be sure you know what happens to the numbers in the stack registers.

Fig. B-1 shows some examples of Hewlett-Packard key strokes which program instructions corresponding to various flow chart statements used in this book. Refer to that figure as you read the following descriptions.

Data is read from the keyboard and automatically is placed in the display register, where it can participate in an arithmetic or function operation or be stored in memory. The machine is stopped to accept data by programming the instruction R/S. This instruction stops the machine if it is running and starts it if it is not. Once the data is in the display, the operator presses R/S to start the machine and the machine then proceeds to the next instruction in the program.

Negative numbers are entered by pressing CHS after the magnitude of the number has been entered. Scientific notation can be used. Following entry of the mantissa, EEX is pressed, followed by the one or two digit power of 10. If the exponent is negative, CHS is pressed immediately after EEX, then the magnitude of the exponent is entered.

Storage locations are labelled by any of the integers 0 through 9 or the letters A, B, C, D, E (the top row of keys). A number in the display is stored in memory location 1, for example, by programming STO 1. Any number previously in that storage location is erased to provide space for the new number. The number placed in storage also remains in display and other numbers in the stack are also unchanged. This means, for example, that a number may be placed in storage and immediately used in an arithmetic operation without first being recalled.

After storage of the first number, a second number may be entered at the keyboard. When this number is entered, the stack lifts: the number previously in X goes to Y, the number previously in Y goes to Z, the number previously in Z goes to T, and the number previously in T is lost. Although we do not do it in this book, it is sometimes desirable to enter the second number without storing the first. In this case, entry of the first number is followed by ENTER. The ENTER instruction lifts the stack and retains the first number in X. It is now in both X and Y. When the second number is entered, it goes into X and the previous content of X is erased. The first number is now in Y and the second is in X.

Figure B-1. Examples of instructions for Hewlett-Packard hand held machines.

a. <u>Enter</u> <u>and</u> <u>store</u>

                ┌─────────────┐
                │   ENTER   \
                │   X1, X2   /
                └─────────────┘

R/S   machine stops, user keys in X1, presses R/S to restart machine.
STO 1
R/S   machine stops, user keys in X2, presses R/S to restart machine.
STO 2

b. <u>Operations</u>

     ┌───────────────┐
     │   5*X1→X2   │   5 RCL 1 × STO 2
     └───────────────┘

     ┌───────────────┐
     │ X1+X2→X3   │   RCL 1 RCL 2 + STO 3
     └───────────────┘

     ┌───────────────┐
     │ (X1+X2)*X3→X4 │   RCL 1 RCL 2 + RCL 3 × STO 4
     └───────────────┘

     ┌───────────────┐
     │ X1+X2/X3→X4   │   RCL 2 RCL 3 ÷ RCL 1 + STO 4
     └───────────────┘

     ┌───────────────────┐
     │ (X1+X2)/(X3+X4)→X5 │   RCL 1 RCL 2 + RCL 3 RCL 4 + ÷ STO 5
     └───────────────────┘

     ┌───────────────┐
     │ SIN(X1)→X2   │   RCL 1 f SIN STO 2
     └───────────────┘

     ┌───────────────┐
     │ X1*SIN(X2*X3) │   RCL 2 RCL 3 × f SIN RCL 1 × STO 4
     │   →X4   │
     └───────────────┘

Figure B-1. Cont'd. Hewlett-Packard.

| INVTAN(X1)→X2 | RCL 1 g tan^{-1} STO 2 |

| X1↑2→X2 | RCL 1 g x^2 STO 2 |

| SQRT(X1↑2 +X2↑2)→X3 | RCL 1 g x^2 RCL 2 g x^2 + f √x STO 3 |

| X1↑1.5→X2 | RCL 1 1.5 h y^x STO 2 |

| f(X1,X2)→X7 | 5 RCL 1 × RCL 2 + STO 7 |
where f(X1,X2) =5*X1+X2

c. <u>Display results</u>

| DISPLAY X1, X2 |

RCL 1 R/S   User copies result, presses R/S to restart machine.

RCL 2 R/S   User copies result, presses R/S to restart machine.

Figure B-1. Cont'd. Hewlett-Packard.

d. <u>Transfer Statements</u>

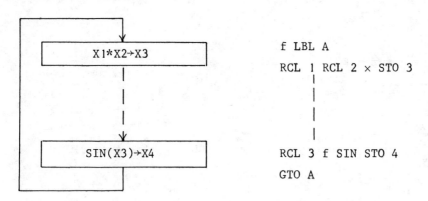

f LBL A
RCL 1 RCL 2 × STO 3
|
|
|
RCL 3 f SIN STO 4
GTO A

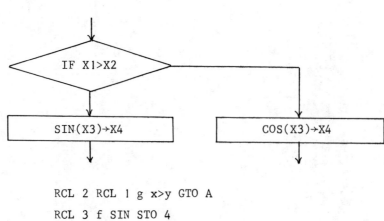

RCL 2 RCL 1 g x>y GTO A
RCL 3 f SIN STO 4
|
| Other statements on the left
| side of the flow chart.
|
f LBL A
RCL 3 f COS STO 4

Figure B-1. Cont'd. Hewlett-Packard.

e. <u>Additional</u> statements

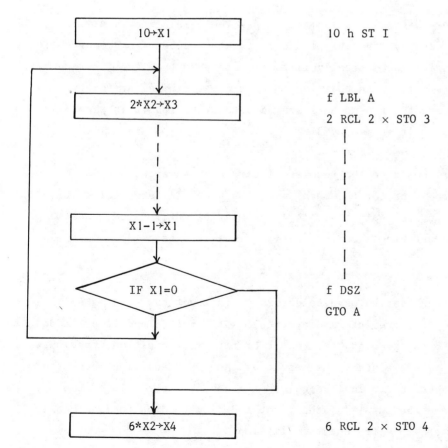

X1+X2→X1   RCL 2 STO + 1

X1*X2→X1   RCL 2 STO × 1

(5+X2↑2)+X1→X1   RCL 2 g x^2 5 + STO + 1

10→X1   10 h ST I

2*X2→X3   f LBL A
                  2 RCL 2 × STO 3

X1-1→X1

IF X1=0   f DSZ
                  GTO A

6*X2→X4   6 RCL 2 × STO 4

A number is recalled to X from memory location 1, for example, by the instruction RCL 1. The number remains in memory as well, and the numbers in the stack move up. An exception to this behavior occurs when the recall follows the execution of an ENTER instruction. The ENTER instruction causes the machine to copy the number in X into Y so that the same number is now in both X and Y. The recall instruction then simply erases the number in X and replaces it with the number from memory.

Arithmetic operations are carried out with the numbers in X and Y. When one of the instructions +, -, ×, ÷, or y^x is executed, the operation is immediately carried out and the result placed in the display. The result is X+Y, X-Y, X×Y, X/Y, or Y^X, depending on the instruction executed. For the last operation, the number in Y must be positive. In all cases, previous instructions must be used to place the correct numbers in X and Y.

When the result is placed in X, the number previously in Y is lost, the number previously in Z goes to Y, and the number previously in T goes to Z and also remains in T. The stack is said to drop.

To accomplish the flow chart line X1+X2→X3, we ask the machine to recall the number from memory 1 to the display, then ask it to recall the number from memory 2 to the display. This second action automatically lifts the stack so that the first number is now in Y. The two numbers are now in position to participate in the arithmetic operation. When + is executed, the two numbers are added and the result is placed in the display. STO 3 then stores it in memory 3.

The key strokes given should not be read as algebraic statements, but as what they are: a series of instructions for the machine to carry out in sequence. RCL 1 RCL 2 +, for example, makes no sense algebraically. It must be interpreted in terms of what the machine does on encountering each instruction. What the stack does must also be taken into account.

The behavior of the stack can be used to carry out instructions such as (X1+X2)/(X3+X4)→X5 without using STO to store intermediate results. First X1 and X2 are added in the usual way. When RCL 3 is executed, the result of the first addition, previously in X, moves to Y. When RCL 4 is executed, the result of the first addition moves to Z, X3 moves to Y, and X4 is placed in X. The execution of + causes the result of the second addition to be in X and the result of the first addition to drop to Y. These numbers are now properly placed to carry out the division.

The machine has a number of function keys. These are

f	LN	natural logarithm of X,
g	e^x	exponential of X,
f	LOG	logarithm to the base 10 of X,
g	10^x	10 to the power X,
f	$\sqrt{x}$	square root of X,
g	x^2	square of X,
h	$1/x$	reciprocal of X,
h	ABS	absolute value of X,
f	SIN	sine of X,
g	SIN^{-1}	arc sine of X,
f	COS	cosine of X,
g	COS^{-1}	arc cosine of X,
f	TAN	tangent of X,

and $\quad$ g TAN^{-1} $\quad$ arc tangent of X.

These operate immediately on the number in the display and replace it with the value of the function. The rest of the stack remains unchanged. The keys h π lift the stack and put the value for π in X.

Some of the function symbols appear above the keys on the top row as well as on other keys, used in conjunction with the f, g, or h keys. In a program, the keys in the top row can be used only to program the letters which appear on them and cannot be used to program the functions. On the other hand, when the machine is run manually, these keys cause it to perform the functions and the letters have no meaning.

Some programs in this book use a function which is defined by the user at the time the machine is programmed. A typical flow chart statement is shown as the last example of part b, Fig. B-1. At the time of programming, the user should substitute instructions which cause the machine to evaluate the function and store its value in the memory location indicated by the flow chart. More advanced programmers should use the subroutine capability of the machine. See the instruction manual for details.

Output is obtained by recalling the number of interest from memory, then executing R/S. This stops the machine so that the number, which is then in the display, can be copied. The operator then presses R/S to restart the machine. It is possible to control the number of digits displayed. See the instruction manual.

Program instructions can be labelled with any of the letters A, B, C, D, or E. The appropriate key strokes to label an instruction A, for example, are f LBL A. It is also possible to use lower case letters (preceded by f) or the integers 0 through 9. The labelling statement precedes the labelled instruction.

Unconditional transfers are accomplished by programming GTO, followed by the label which identifies the instruction to be executed next. For purposes of the example given in Fig. B-1, we assume we want the machine to return to instruction A after it completes the last instruction shown.

A conditional transfer starts with either a comparison of the number in X and the number in Y or else a comparison of the number in X with 0. One possible example uses the g x>y keys. If the comparison statement is true (if the number in X is greater than the number in Y), then the machine executes the next instruction in the program. This might be a GTO instruction. If, on the other hand, the comparison statement is false (if the number in X is less than or equal to the number in Y), then the machine ignores the next instruction and next executes the instruction after that one.

Other comparison statements available are

	g x≤y	true is X is less than or equal to Y,
	g x≠y	true if X is not equal to Y,
	g x=y	true if X is equal to Y,
	f x<0	true if X is less than zero,
	f x>0	true if X is greater than zero,
	f x≠0	true if X is not equal to zero,
and	f x=0	true if X is equal to zero.

The last four represent a comparison of the number in X with zero. To use the first group, for which the numbers in X and Y are compared, care must be taken to see that the proper numbers have been placed in the X and Y registers. The keys h x⇄y are sometimes useful in this regard. They interchange the contents of X and Y.

It is possible to perform some arithmetic operations directly in memory, rather than recalling both partners from memory. The appropriate instructions, for memory 1, are STO + 1, STO - 1, STO × 1, and STO ÷ 1. In the first case, the display number is added to the number in memory 1; in the second case, the display number is subtracted from the number in memory 1; in the third case the number in memory 1 is multiplied by the display number; and, in the last case, the number in memory 1 is divided by the display number. In all cases, the result is placed in memory 1 and the stack is uneffected. Other memory locations, of course, can be used by replacing the 1 in the above instructions by another memory location label, but only memory locations 0 through 9 can be used. Arithmetic operations cannot be performed in the lettered storage locations.

The machine allows for a special way of counting the number of times it traverses a loop. This instruction is not really necessary since the same result can be produced by instructions already given, but its use decreases running time so much that it is worthwhile discussing it. First, outside the loop, the number of times the loop is to be traversed is placed in a special storage location, called the I register. This is accomplished by first placing the number in the display and then executing h ST I. The loop statements are then given. The first statement in the loop is labelled, with an A, for example, the next to last statement is f Dsz, and the last statement is GTO A (the label for the first statement of the loop).

When the machine encounters the DSZ instruction, it reduces the number in I by 1, then checks to see if the result is zero. If it is not, the machine then executes GTO A and returns to the beginning of the loop. If it is, the machine ignores the GTO A instruction and executes the first instruction following the loop. An example is given in Fig. B-1.

Fig. B-2 gives the key strokes for the program described in the introduction to the appendices. This program takes into account two important features of Hewlett-Packard machines. The first line is a label. When the program is run, the user, after placing the W/PRGM-RUN switch in the RUN position, presses the key marked A, the label of the first line. The machine then goes to instruction A and begins execution. In this case, of course, it immediately stops to receive X1.

The second feature is found in the last line of the program. The instruction is h RTN. The machine, when it encounters this instruction, stops. When the user presses A, it goes back to the beginning of the program and begins execution there.

Figure B-2. Sample program.

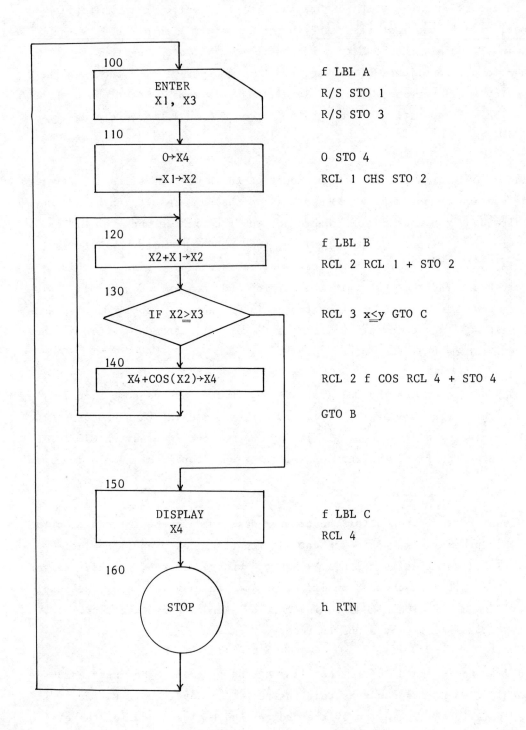

All programs should begin with a label to allow the user to reset the machine to the first instruction. They should all end with h RTN, even if this instruction is never encountered when the program is run. The machine, when it searches for a label, starts with the current instruction, a transfer instruction, and looks at later instructions, in sequence. When it reaches h RTN without finding the label, it goes to the beginning of the program and continues the search. For the program of Fig. B-2, the machine would not find instruction B were the h RTN not there. This is because instruction B precedes the transfer statement, at line 130, where the search commences.

The rest of the keystrokes are straightforward. You should carefully compare them to the flow chart statements.

Some of the programs in this book require that the memory location number be computed within the program. For example, the variables X(3*X2+1), X(3*X2+2), and X(3*X2+3) might appear in flow chart statements. The X's are called subscripted variables and the numbers or expressions in the parentheses are called subscripts. They are the memory location numbers.

Subscripted variables are handled as follows. First, the subscript is calculated and placed in the special memory I. Storage is accomplished by means of the instruction h ST I. For the first subscripted variable above, the correct sequence of key strokes is

$$3 \; RCL \; 2 \; \times \; 1 \; + \; h \; ST \; I$$

If the number X(3*X2+1) is in the display, it can be stored in 3*X2+1 by means of STO (i). X(3*X2+1) can be recalled from memory to the display by means of the keys RCL (i).

The value in I can be any integer 0 through 25. The first 10 correspond to the usual storage locations 0 through 9. The next 10 correspond to a special set of 10 locations we have not yet mentioned. For our purposes here, they may be used just as the other locations are used. They cannot be used, as the other locations are, for ordinary storage and recall and the interested reader is referred to the instruction manual for details. The next 5 integers (20 through 24) correspond to the lettered locations A, B, C, D, and E. Finally, 25 corresponds to I itself. None of our programs require putting the integer 25 into I. Thus, when subscripted variables are used, the subscript can take on any of the 25 values, 0 through 24.

Appendix C
BASIC

There are many versions of BASIC language, differing slightly from each other, and the reader should consult the instruction manual for the machine to be used. We give one popular version here. The steps needed to prepare the machine for programming and running also differ from installation to installation.

In BASIC, all variables are named with a one or two symbol name. The first symbol is a letter, the second may be either a letter or a number. In keeping with our flow chart convention, we use X1, X2, ... X9. If more than 9 variables are needed use Y1, Y2, ... These names designate the memory locations where the values are stored.

All statements are numbered by the user and the number of a statement is the first entry for that statement. To remind you, all our sample statements are numbered in the following examples. It is wise to number by 10's so that additional statements can be inserted later. No two statements may have the same number.

The last statement of any program is END to let the machine know that there are no more statements. Some machines have other signals.

Fig. C-1 shows some samples of statements written in BASIC. They are similar to those which appear in the flow charts of this book.

Entry of data can be accomplished in one of two ways. If entry is to be from the keyboard, the statement INPUT X1, X2 is used. The list of variables can be as long as you like, with the variables separated by commas. When the machine encounters this statement, it stops. The user then keys in the number for X1 and presses RETURN. The machine accepts the number and stops to await X2. When return is pressed after the last entry, the machine proceeds to the next statement. The entries are automatically stored in memory locations set aside for them. The number 100, which precedes the word INPUT in the example, is a statement number, and may be replaced by any other convenient number, less than some large number which depends on the installation.

Assignment of values to the variables can also be accomplished by means of statements within the program. The two statements READ X1, X2 and DATA 5.3,7.1 place

Figure C-1. Examples of instructions for BASIC.

a. Enter and store

```
         ┌─────────────┐
         │   ENTER     \
         │   X1, X2     \
         └───────────────┘
```

```
100 INPUT X1,X2
        or
100 READ X1,X2
110 DATA 5.3,7.1
```

b. Operations

```
┌─────────────┐
│   5*X1→X2   │
└─────────────┘
```
100 X2=5*X1

```
┌─────────────┐
│  X1+X2→X3   │
└─────────────┘
```
100 X3=X1+X2

```
┌──────────────────┐
│ (X1+X2)*X3→X4    │
└──────────────────┘
```
100 X4=(X1+X2)*X3

```
┌─────────────┐
│  SIN(X1)→X2 │
└─────────────┘
```
100 X2=SIN(X1)

```
┌──────────────────┐
│  X1*SIN(X2*X3)   │
│       →X4        │
└──────────────────┘
```
100 X4=X1*SIN(X2*X3)

```
┌────────────────┐
│ INVTAN(X1)→X2  │
└────────────────┘
```
100 X2=ATAN(X1)

```
┌─────────────┐
│   X1↑2→X2   │
└─────────────┘
```
100 X2=X1↑2

Figure C-1. Cont'd. BASIC.

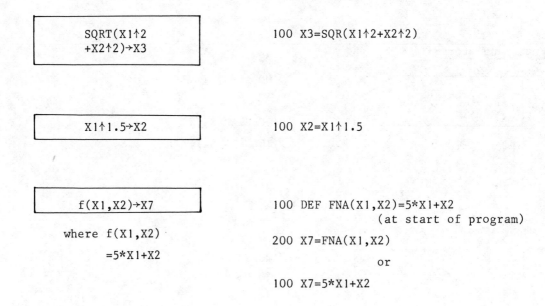

c. Display results

	DISPLAY X1, X2	100 PRINT X1,X2

d. Transfer statements

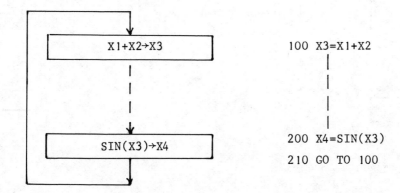

```
100 X3=X1+X2
      |
      |
      |
      |
200 X4=SIN(X3)
210 GO TO 100
```

Figure C-1. Cont'd. BASIC.

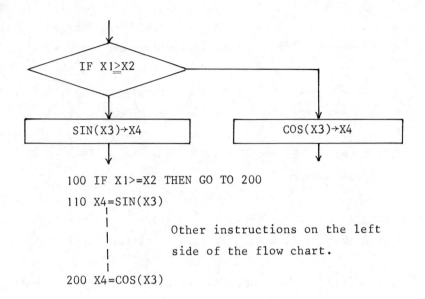

```
100 IF X1>=X2 THEN GO TO 200
110 X4=SIN(X3)
    |
    |   Other instructions on the left
    |   side of the flow chart.
    |
200 X4=COS(X3)
```

e. <u>Additional</u> <u>statement</u>

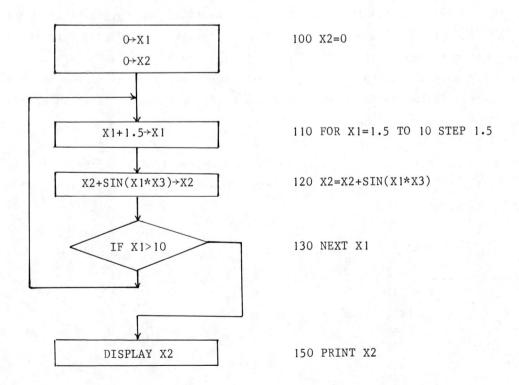

```
100 X2=0

110 FOR X1=1.5 TO 10 STEP 1.5

120 X2=X2+SIN(X1*X3)

130 NEXT X1

150 PRINT X2
```

5.3 in X1 and 7.1 in X2. The scheme can be extended to a longer list of variables. DATA statements may appear anywhere in the program, but they are usually placed after the READ statement which calls for them or else they are placed together at the end of the program. They are read in order of increasing statement number.

For INPUT entry, numbers are entered by pressing appropriate keys, preceded by the sign if the number is negative. Scientific notation may be used. Following the mantissa, the letter E is entered, then the sign of the exponent and the one or two digit power of 10. Positive signs may be omitted. The same format is used when the data is entered within the program with the READ statement.

Arithmetic operations are written in the form X3 = X1 + X2. This means the number in X1 is added to the number in X2 and the result is stored in X3. The symbol = does not necessarily mean "equals" in the algebraic sense. X1 = X1 + X2 is a valid statement for any value of X1 or X2. Other operations are - (subtract), * (multiply), / (divide), and ↑ (exponentiate). Examples are given in Fig. C-1.

A sequence of arithmetic operations may be strung together as in

$$100 \quad X1 = X2 * X3 + X5 / X6$$

Multiplications and divisions are carried out before additions and subtractions so X2 is multiplied by X3, then X5 is divided by X6 and the two results are added. Exponentiation takes precedence over the other operations, so X1*X2↑1.5 means that X2 is raised to the 1.5 power and the result is multiplied by X1. Parentheses may be used freely and they should be used to eliminate any doubt about the order of operations. Inner parentheses are evaluated first, then outer ones.

The arguments of a function are written in parentheses following the function name. Thus X2 = SIN(X1) means find the sine of the number in X1 and place it in X2. Other common functions are

        COS(X1)   cosine of X1,
        ATN(X1)   arc tangent of X1,
        SQR(X1)   square root of X1,
        ABS(X1)   absolute value of X1,

	EXP(X1)	exponential of X1,
	LOG(X1)	natural logarithm of X1,
	INT(X1)	the greatest integer less than or equal to X1,
and	SGN(X1)	+1 if X1 is positive, -1 if X1 is negative.

Many machines allow other functions.

The argument of a function may be any valid BASIC expression which can be evaluated as a number at the time the function is encountered. The machine first evaluates the expression, then the function. X2 = EXP(2*X1+5) is a valid statement provided there is a number stored in X1 when it is encountered.

BASIC allows for user defined functions. The statement used to define a function has the form

    100   DEF FNA(X1,X2) =

followed by an expression for the function in terms of X1 and X2. For example, if the function to be evaluated is f(x,y)=x sin(y), then the complete statement is

    100   DEF FNA(X1,X2) = X1 * SIN(X2)

This statement appears at the beginning of the program. Later, the function may be used in a statement such as, for example,

    200   X5 = X2 * FNA(2*X3,X4)

The variables X1 and X2 in the defining statement are dummy variables and, in use, may be replaced by any valid BASIC expressions which can be evaluated as numbers. In the above example, 2*X3 is evaluated and the result used wherever X1 is used in the defining statement. X4 is used wherever X2 is used in the defining statement.

Any number of variables may appear in the argument list, with the variables in the list separated by commas. If more than one function is used in the same program, they are named FNA, FNB, FNC, etc. Actually, the third letter may be any letter of the alphabet.

The PRINT instruction, followed by a list of variables, causes the values of the variables named to be displayed on the cathode ray tube face or printed on paper, depending on the output device used.

As shown in part d of Fig. C-1, user assigned statement numbers identify statements for purposes of transfer instructions. The meaning of

$$200 \quad \text{GO TO} \quad 100$$

is obvious. The machine, when it encounters this statement, immediately goes to statement 100, wherever it is, and begins executing instructions there.

Conditional transfer statements have the form

$$100 \text{ IF } X1<X2 \text{ THEN GO TO } 200$$
$$110 \text{ } X2 = 3 * X5$$

If the value of X1 is less than that of X2, then the comparison statement is true and the machine proceeds to statement 200. If the value of X1 is greater than or equal to that of X2, then the comparison statement is false and the machine immediately goes to the next statement, statement 110, in this case.

Other comparison statements which can be used are

	X1=X2	true if X1 equals X2,
	X1>X2	true if X1 is greater than X2,
	X1>=X2	true if X1 is greater than or equal to X2,
	X1<=X2	true if X1 is less than or equal to X2,
and	X1<>X2	true if X1 is not equal to X2.

Some BASIC systems allow the use of two special statements to construct loops in which one variable is incremented or decremented. The first statement of the loop might be, for example,

$$100 \text{ FOR } X = 1 \text{ to } 3 \text{ STEP } .2$$

and the last statement of the loop is

$$150 \text{ NEXT } X$$

The first time through the loop, the variable X is assigned the value 1 and each successive time around the loop, it is incremented by 0.2 until its value is greater than 3. When this occurs, the machine proceeds to the instruction which follows the NEXT X statement. The incrementation takes place at the last line, NEXT X, and the check to see if X is larger than 3 occurs immediately after X has been incremented.

The variable used, the initial and maximum values of the variable, and the step size are all arbitrary. In fact, the initial and maximum values of the variable, and the step size, may be written as BASIC expressions, which are evaluated at the time of execution.

The use of these statements is shown in part e of Fig. C-1. Note that X1 is never equal to 10 for the example given. The last time through the loop, X1 has the value 9. If the first statement of the loop were 110 FOR X1=1 TO 10 STEP 1, then the last time through the loop X1 would be 10.

Fig. C-2 shows a flow chart for the sample program described in the introduction to the appendices. Instructions in BASIC are given beside the flow chart statements. Compare carefully.

A few of the programs in this book use variables whose identifying numbers or storage locations must be computed by the machine during the running of the program. For example, suppose we wish to store numbers in X(3*X2+1), X(3*X2+2), and X(3*X2+3), where X2 is a variable and takes on the values 1, 2, 3, ... in a loop. The numbers in parentheses are the memory location numbers. The X's are called subscripted variables and the numbers in the parentheses are called subscripts.

The notation used above is valid in BASIC. That is X (expression which can be evaluated) can be used to designate a variable and used in BASIC statements. The following are three sample uses of the first of the variables mentioned above:

```
            100 X1 = 5 * X(3*X2+1)
            200 PRINT X(3*X2+1)
and         300 INPUT X(3*X2+1)
```

These are all valid BASIC statements provided there is a number stored in X2 so that the machine can evaluate the expression in the parentheses.

Figure C-2. Sample program.

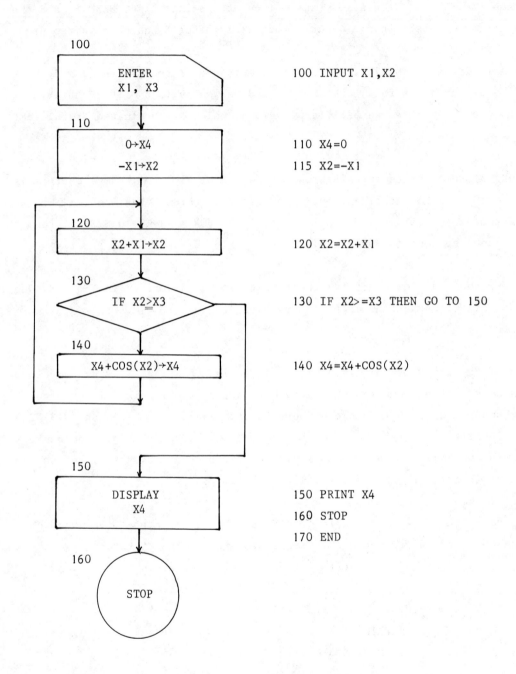

One other statement is needed when subscripted variables are used. It is called a dimension statement and it is placed at the beginning of the program, before the first executable statement of the program and before any function definition statements. It tells the machine how many storage spaces to set aside for the subscripted variables and it has the form, for example

        100   DIM X(18)

where the number in parentheses, called the dimension of X, is the maximum value of the subscript. The dimension given can be larger than the largest subscript actually used, but, if a dimension is given which is smaller than the largest subscript used, the program probably will not run properly.

APPENDIX D
FORTRAN

FORTRAN differs slightly from machine to machine. The biggest differences are in the input and output statements, but other statements may be different as well. We concentrate on the case when the instructions and data are entered into the machine by means of punched cards. The reader should check the instruction manual for other methods of input, for instructions to prepare the machine to accept and run a program, and for slight differences in the FORTRAN statements themselves.

Fig. D-1 gives some examples of various types of FORTRAN statements, keyed to the flow chart statements of this book. Refer to it as you read the following description of the FORTRAN language.

Variables in FORTRAN are named with up to 5 symbols, the first of which must be a letter. The others may be letters or numbers. We use X1, X2, etc., in keeping with the symbols used in the flow charts of this book. If the name of the variable begins with I, J, K, L, M, or N, the number is called a fixed point number. It must be a positive or negative integer, the machine does not carry the decimal point, and scientific notation cannot be used. For other variables, called floating point variables, the decimal point is carried and scientific notation can be used. Most of the variables we use are floating point, but a few of the instructions require fixed point variables. Variable names identify storage locations.

The cards read by the machine may contain FORTRAN statements or numerical data. There are 72 positions in a FORTRAN statement line and all statements which fit on a single card may consist of up to 72 symbols. The first 5 positions are reserved for statement numbers, although all are not normally used. In fact, statements need not be numbered unless they are referred to by the program. The sixth position is normally left blank and the others are used to write the instruction. If more positions are needed for an instruction, it may be continued onto succeeding cards by placing a symbol, which is not part of the instruction, in position 6 of the following cards used for the continuation. There can be no more than one statement on a card.

For cards which contain numerical data, the data may start in the first position and continue across the card. While statement cards do not use the last eight positions,

Figure D-1. Examples of FORTRAN instructions.

a. <u>Enter and store</u>

```
┌──────────────┐
│   ENTER      │
│   X1, X2     │
└──────────────┘
```

||READ(5,20)X1,X2
20||FORMAT(E14.8,E14.8)
    or
||READ,X1,X2

b. <u>Operations</u>

```
┌──────────────┐
│   5*X1→X2    │
└──────────────┘
```
||X2=5.*X1

```
┌──────────────┐
│   X1+X2→X3   │
└──────────────┘
```
||X3=X1+X2

```
┌──────────────┐
│ (X1+X2)*X3→X4│
└──────────────┘
```
||X4=(X1+X2)*X3

```
┌──────────────┐
│   SIN(X1)→X2 │
└──────────────┘
```
||X2=SIN(X1)

```
┌──────────────┐
│ X1*SIN(X2*X3)│
│      →X4     │
└──────────────┘
```
||X4=X1*SIN(X2*X3)

```
┌──────────────┐
│ INVTAN(X1)→X2│
└──────────────┘
```
||X2=ATAN(X1)

```
┌──────────────┐
│   X1↑2→X2    │
└──────────────┘
```
||X2=X1**2

Figure D-1. Cont'd. FORTRAN.

```
┌─────────────────┐
│  SQRT(X1↑2      │         ||X3=SQRT(X1**2+X2**2)
│   +X2↑2)→X3     │
└─────────────────┘


┌─────────────────┐
│   X1↑1.5→X2     │         ||X2=X1**1.5
└─────────────────┘


┌─────────────────┐             ||AF(X1,X2) =5.*X1+X2
│   f(X1,X2)→X7   │                    (Before all executable
└─────────────────┘                     statements.)
   where f(X1,X2)             ||X7=AF(X1,X2)
     =5*X1+X2                          or
                              ||X7=5.*X1+X2
```

c. <u>Display</u> or <u>print</u> <u>results</u>

```
┌─────────────────┐         ||WRITE(6,50)X1,X2
│     DISPLAY     │      50||FORMAT(E14.8,E14.8)
│      X1, X2     │                or
└─────────────────┘         ||PRINT,X1,X2
```

d. <u>Transfer</u> <u>statements</u>

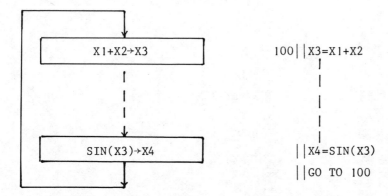

Figure D-1. Cont'd. FORTRAN.

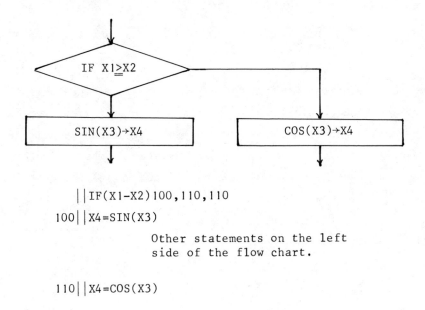

```
     ||IF(X1-X2)100,110,110
100||X4=SIN(X3)
          Other statements on the left
          side of the flow chart.

110||X4=COS(X3)
```

e. <u>Additional</u> <u>statement</u>

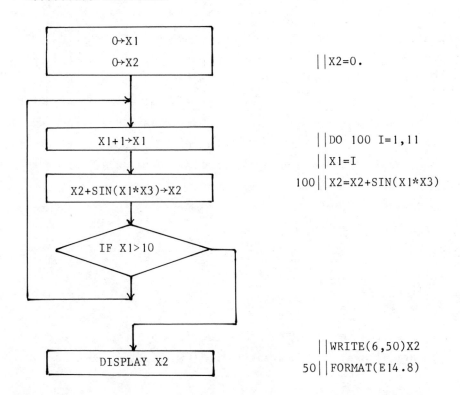

```
     ||X2=0.


     ||DO 100 I=1,11
     ||X1=I
100||X2=X2+SIN(X1*X3)




     ||WRITE(6,50)X2
 50||FORMAT(E14.8)
```

data may be placed in all eighty positions of the card.

The usual statement used to enter data has the form

||READ(5,50)X1,X2,X3

where the list following the second parenthesis contains the names of the variables to be read, separated by commas. The double line in this and other statements represents the sixth position on the card. The first number inside the parentheses specifies the source of the data. 5 usually stands for a card reader and other numbers denote other sources, such as a keyboard or tape reader. The second number, 50 in this case, refers to a format statement, which tells how the numbers are distributed on the data card.

Although many other formats are available,

50||FORMAT(E14.8,E14.8,E14.8)

does the job for the list in the READ statement above. The number 50 is the statement number referred to by the READ statement. E14.8 means that the number is given in 14 spaces on the card. These spaces are: the sign of the number, a decimal point, 8 digits of the mantissa, the letter E (to indicate exponent), the sign of the exponent, and 2 numbers to give the magnitude of the exponent of 10. An example of an E14.8 number is $-.17327156E-04$, which means $-.17327156 \times 10^{-4}$. The second number is given in positions 15 through 28 of the card, and the third number is given in positions 29 through 52. The number may take fewer than 14 spaces if positive signs are omitted or fewer significant figures are written. It is admissible then to write the number anywhere within the 14 spaces allotted to it. Nevertheless, the second and other numbers must be written in the spaces allotted to them, as if they each used 14 symbols. Data cards follow the program and are placed in the order in which they are read. There is one data card for each READ statement.

Arithmetic statements have the form

||X1=X2+X3

which means: add the number stored in X2 to the number stored in X3 and store the result in X1. Other possible operations are: - (subtraction), * (multiplication),

/ (division), and ** (exponentiation).

Operations may be strung together and, in any sequence, exponentiation is carried out first, then multiplication and division, and finally addition and subtraction. In

$$X1=X2+X3*X4**2$$

X4 is squared and the result is multiplied by X3. That result is then added to X2.

The expression to the right of the = symbol cannot contain both fixed and floating point numbers. The only exception to this rule is that an exponent may be fixed point, as is the 2 in the above expression. Note carefully the decimal points in the examples of Fig. D-1.

Parentheses may be used to designate the order of operation. The content of the inner most set is evaluated first, then the content of the next inner most, and so on until the content of the outer set is evaluated.

FORTRAN statements should not be read as algebraic expressions. For example, the statement X1=X1*X2+X3 is valid no matter what the current values of X1, X2, and X3. The statement means: multiply the number in X1 by the number in X2, add the result to the number in X3, and store the answer in X1, thereby erasing the previous content of X1.

Functions are identified by a name followed by an argument, placed inside a set of parentheses. The argument may be a floating point number, a floating point variable, or a floating point expression which can be evaluated by the machine.
Thus

$$X1=SIN(X2)$$

tells the machine to compute the sine of the number in X2 and store the result in X1. Most machines have available a large library of FORTRAN functions. Some common ones are

COS(X1)	cosine of X1,
TAN(X1)	tangent of X1,
ASIN(X1)	arc sine of X1,
ATAN(X1)	arc tangent of X1,
EXP(X1)	exponential of X1,

	LOG(X1)	natural logarithm of X1,
	SQRT(X1)	square root of X1,
and	ABS(X1)	absolute value of X1.

Examples of the use of arithmetic operations and functions are given in Fig. D-1.

Some of the programs in this book contain unspecified functions, which must be supplied by the user when the machine is programmed. This can be done by substituting the FORTRAN statements which cause the machine to evaluate the function and store its value in the memory location given in the flow chart. For example, suppose the flow chart statement is f(X1,X2)→X3 and, for a particular problem, suppose the function is 3*X1+X2. Then, the FORTRAN line is

$$||X3=3.*X1+X2$$

FORTRAN allows a second method to be used. Before all executable statements in the program, a function can be named and defined. If we call the function AF, then the defining statement is

$$||AF(X1,X2)=3.*X1+X2$$

for the function described above. At any point in the program, the function can be used by writing it in a statement. For example,

$$||X3=5.*AF(X1,X2)$$

evaluates the function, multiplies its value by 5, then stores the result in X3.

The number of variables in the list of arguments is arbitrary. If there is more than one, their names are separated by commas. When called by the program, the arguments need not be the same as those used in the defining statement.

$$||X3=AF(5.,6.*X2)$$

is a valid FORTRAN statement. 5. is substituted for X1 where it occurs in the defining statement and the value of 6.*X2 is substituted for X2 where it occurs.

Output is obtained by means of a statement such as

$$\text{WRITE}(6,20)X1,X2,X3$$

where the list of variables may be any convenient length and the names of the variables are separated by commas. The first number inside the parentheses denotes the output device. 6 usually refers to an on line printer and other devices have other identifying numbers. The number 20 refers to a format statement, such as

$$20\ \text{FORMAT}(E14.8,E14.8,E14.8)$$

This has the same meaning as the format statement discussed in connection with READ statements. Now it describes how the numbers are written by the output device.

For some installations, input and output of data can take place without specification of format or device, some standard device being used. The statements have the form

$$\text{READ},X1,X2,X3 \qquad \text{for input}$$

and

$$\text{PRINT},X1,X2,X3 \qquad \text{for output.}$$

The input data card then gives 3 numbers (one for each variable on the list), separated by commas. This form may also be used when input is via a keyboard. Output is in some standard form, such as E14.8 and is usually printed by an on line printer.

Examples of FORTRAN transfer statements are given in part d of Fig. D-1. An unconditional transfer statement has the form

$$\text{GO TO } 100$$

where 100 is the statement number of the statement to be executed next.

Conditional transfer is accomplished by a statement of the form

$$\text{IF}(X1)100,200,300$$

where the numbers are statement numbers. The machine proceeds next to statement 100 if X1 is negative, to statement 200 if X1 is zero, and to statement 300 if X1 is positive. Two of the statement numbers may be the same. For example, if the machine is to go to statement 200 for X1 greater than or equal to zero, then the transfer statement should be written

```
IF(X1)100,200,200
```

The quantity in the parentheses after IF may be any FORTRAN expression which can be evaluated as a number. Since it is usual to compare two variables, the expression is usually the difference between the two. Carefully note the example given in Fig. D-1.

The FORTRAN language contains a statement which facilitates the programming of loops when the number of times the loop is to be traversed is known. The statement has the form

```
DO 100 I=I1,I2,I3
```

All of the statements following this one, up to and including the statement numbered 100, are in the loop. They are executed with the fixed point variable I set equal to I1. Then I is incremented by I3 and the loop traversed again. This process is repeated until I, if incremented, would have a value greater than I2. The machine then goes to the statement following statement 100 and begins executing instructions there. Note that the last time through the loop, I has the value I2 or a value slightly less.

I1, I2, and I3 may be replaced in the statement by constant integers as in the example of Fig. D-1. This is the usual way DO statements are written. Statement numbers other than 100 may, of course, be used. I3 may be omitted if the loop variable I is incremented by 1 each time around. In this case, the statement is

```
DO 100 I=I1,I2
```

In the example of Fig. D-1, the content of I must be placed in X1 and thereby converted to a floating point number in order to be used in the next statement. Otherwise, the next statement would be prohibited since it would contain an expression with both fixed and floating point variables.

There may be a transfer statement within the loop and control may be passed to another statement either inside or outside the loop in the usual way. Except in special circumstances, control cannot be passed to a statement inside the loop from a statement outside the loop.

In some programming situations, it is necessary to end a loop with a statement which does nothing. Its sole purpose is to supply a statement to which control may be passed by a transfer statement within the loop. The statement used for this purpose is

        100| |CONTINUE

Fig. D-2 gives a flow chart and FORTRAN statements for the program described in the introduction to the appendices. Carefully compare flow chart and FORTRAN statements. All FORTRAN programs must end with the statement

        | |END

and this one is no exception. It is not an executable statement in the program. Instead, its function is to tell the machine that the last statement has been read and that it can begin processing the program. The END statement should not be confused with the STOP statement, which is a program statement and which, when encountered during the running of the program, causes the machine to halt.

For a few programs in this book, the identifying numbers for some of the variables are computed. In flow charts, we use variables denoted by $X(3*X2+1)$, $X(3*X2+2)$, and $X(3*X2+3)$. These are called subscripted variables and the numbers inside the parentheses are called the subscripts.

The FORTRAN language allows for subscripted variables, but the subscripts must be fixed point numbers. As an example, we suppose $X(3*X2+1)$ is to be multiplied by $X3$ and the result stored in $X7$. The FORTRAN statements which do the job are

        | |I=3*X2+1
        | |X7=X3*X(I)

The first statement gives instructions to calculate the subscript and convert it to fixed point form. This is done by storing it in a location reserved for a fixed point variable. The subscripted variable is then used in a FORTRAN expression just as any other variable is used.

Figure D-2. Sample program.

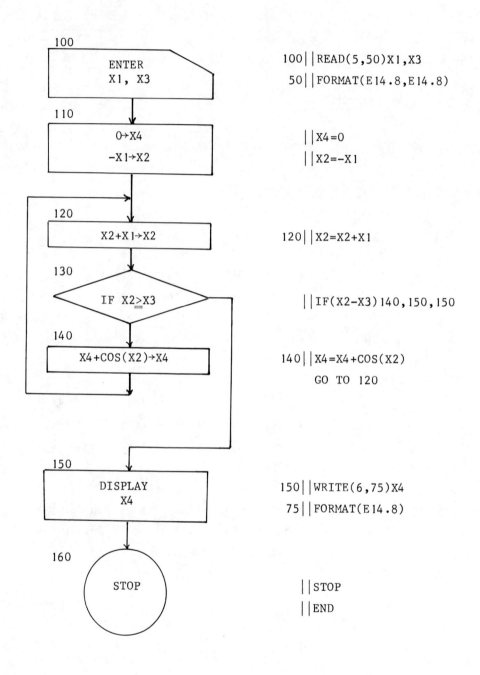

When a subscripted variable is used in a program, one additional statement, called a dimension statement, is required. It has the form

||DIMENSION X(30)

and it must appear before any executable statements, at the beginning of the program. The number between the parentheses tells the machine the largest value possibly assumed by the subscript during the running of the program. No harm results if the number is slightly larger than necessary but, if it is smaller than the maximum value actually attained during running, the program probably will not run correctly.

Appendix E

PASCAL

Pascal statements differ slightly from installation to installation and the reader is encouraged to consult the instruction manual for details. In addition, some specific instructions are necessary to prepare the machine to accept and run a PASCAL program and data. These instructions also vary from installation to installation.

PASCAL is conveniently used with both card and keyboard input. For the description given here, we envisage a card system, but our description can easily be converted. One card corresponds to a typewritten line.

In PASCAL, variables are named with up to 5 symbols. The first must be a letter and the others may be either letters or numbers. We use X1, X2, X3, etc., in keeping with the notation of our flow charts. Variables are classified as either real or integer. If a variable is of type integer, no decimal point is carried by the machine and scientific notation cannot be used. If the variable is of type real, the decimal point is carried and scientific notation can be used. Real variables may be integers but integer variables must be integers. We use integer variables only when required by specific PASCAL statements. Then we use I, J, K, or one of the X's. The variable name specifies the storage location used for the value.

Programs must be given names and the first card names the program. The statement has the form

```
PROGRAM ALPHA(INPUT,OUTPUT);
```

where ALPHA is the name we picked. The words between the parentheses indicate that the standard input device, a card reader say, and the standard output device, a printer say, are used by the program.

Next, the names of all variables used by the program are listed and their types declared. The statement has the form

```
VAR X1,X2,X3,X4,X5,X6:REAL;
    I,J,K:INTEGER;
```

Notice the punctuation for this and the first statement. Variables are separated by commas; a colon precedes the variable type designation; and a semicolon ends the line. In PASCAL, semicolons are used to separate statements and these have been included in anticipation that these statements are followed by others. There may be other declarative statements at the beginning of the program. We discuss these later.

PASCAL statements may start anywhere on the card. The indentation above has been done for ease in reading by humans, not for the machine.

Fig. E-1 shows some examples of PASCAL statements. Refer to them as you read the following descriptions.

Data input takes place via a statement of the form

$$READ(X1,X2,X3)$$

which instructs the machine to read three numbers from a data card and place them in X1, X2, and X3 respectively. Data cards follow the program (or, perhaps an end of record card for some installations), placed in the order they are read by the machine. Each contains numbers to be read, in order. The numbers on any single card are separated by commas and may be placed with any amount of data on a card. It is usual, however, to place all numbers required by any one READ statement on the same card. The data card corresponding to the above READ statement contains three numbers, separated by commas, corresponding to X1, X2, and X3 in that order. Values for real variables may be given in the form -1.735 or in the form -17.35E-1, for example. Here the power of 10 follows the letter E.

Arithmetic operations have the form

$$X1:=X2+X3$$

Notice the colon and the equal sign. This statement means: add the content of X2 to the content of X3 and store the result in X1, erasing the previous content of X1.

Other arithmetic operations are - (subtraction), * (multiplication), and / (division). The variables used may be real or integer and the two types may be mixed in a single expression. If they are mixed, the result is real. Otherwise, the result

Figure E-1. Examples of instructions in PASCAL.

a. <u>Enter and store</u>

```
┌──────────────┐
│   ENTER      │
│   X1, X2     │
└──────────────┘
```
READ(X1,X2)

-5.6,4.2E9   (After all program cards.)

b. <u>Operations</u>

```
┌──────────────┐
│   5*X1→X2    │
└──────────────┘
```
X2:=5*X1

```
┌──────────────┐
│   X1+X2→X3   │
└──────────────┘
```
X3:=X1+X2

```
┌──────────────┐
│ (X1+X2)*X3→X4│
└──────────────┘
```
X4:=(X1+X2)*X3

```
┌──────────────┐
│   SIN(X1)→X2 │
└──────────────┘
```
X2:=SIN(X1)

```
┌──────────────┐
│ X1*SIN(X2*X3)│
│      →X4     │
└──────────────┘
```
X4:=X1*SIN(X2*X3)

```
┌──────────────┐
│ INVTAN(X1)→X2│
└──────────────┘
```
X2:=ARCTAN(X1)

```
┌──────────────┐
│   X1↑2→X2    │
└──────────────┘
```
X2:=SQR(X1)

```
┌──────────────┐
│  SQRT(X1↑2   │
│  +X2↑2)→X3   │
└──────────────┘
```
X3:=SQRT(SQR(X1)+SQR(X2))

```
┌──────────────┐
│  X1↑1.5→X2   │
└──────────────┘
```
X2:=EXP(1.5*LN(X1))

Figure E-1. Cont'd. PASCAL.

```
    ┌─────────────────┐
    │  f(X1,X2)→X7    │
    └─────────────────┘
     where f(X1,X2)
         =5*X1+X2
```

```
X7:=5*X1+X2
    or
PROCEDURE FUNC(X1,X2:REAL;VAR X7:REAL);
    BEGIN
        X7:=5*X1+X2
    END;
(At end of declaration statements.)
FUNC(X1,X2,X7)          (In program.)
```

c. <u>Display</u> or <u>print</u> <u>results</u>

```
    ┌─────────────────┐
    │     DISPLAY     │
    │     X1, X2      │
    └─────────────────┘
```

WRITELN(X1,X2)

d. Transfer <u>statements</u>

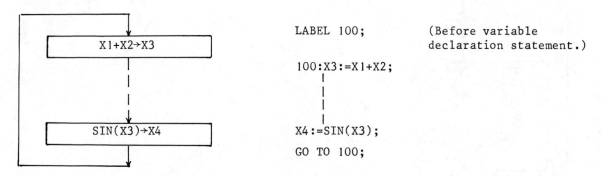

```
LABEL 100;          (Before variable
                     declaration statement.)
100:X3:=X1+X2;
    |
    |
    |
X4:=SIN(X3);
GO TO 100;
```

Figure E-1. Cont'd. PASCAL.

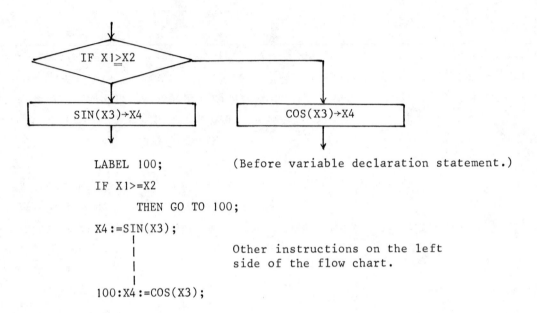

```
LABEL 100;            (Before variable declaration statement.)
IF X1>=X2
    THEN GO TO 100;
X4:=SIN(X3);
    |
    |                 Other instructions on the left
    |                 side of the flow chart.
    |
100:X4:=COS(X3);
```

e. <u>Additional instruction</u>

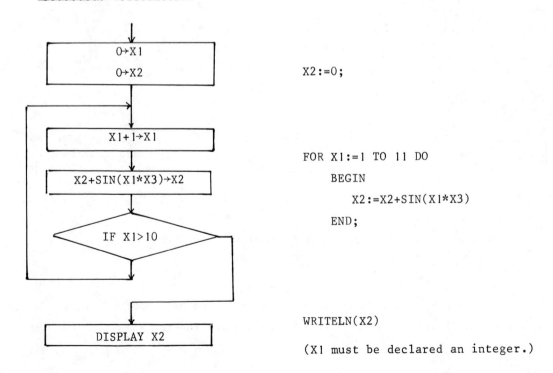

```
X2:=0;



FOR X1:=1 TO 11 DO
    BEGIN
        X2:=X2+SIN(X1*X3)
    END;



WRITELN(X2)

(X1 must be declared an integer.)
```

is the same type as the operands. There is another operation, denoted by DIV. Here the variables must be integer and the result, which is computed by dividing the first number by the second, is also integer. The result is truncated if necessary.

Arithmetic operations may be strung together and parentheses may be used freely to indicate the order in which the operations are to be performed. Inner parentheses are evaluated first, then outer ones. Within each set of parentheses and for an expression as a whole, multiplications and divisions are carried out first, then additions and subtractions. For the statement

$$X1:=X2+X3*X4$$

the content of X3 is multiplied by the content of X4 and the result is added to the content of X2.

There are a number of functions available in PASCAL. Common ones are

ABS(X1)	absolute value of X1,
SIN(X1)	sine of X1,
COS(X1)	cosine of X1,
ARCTAN(X1)	arc tangent of X1,
EXP(X1)	exponential of X1,
LN(X1)	natural logarithm of X1,
SQR(X1)	square of X1,
SQRT(X1)	square root of X1,

and TRUNC(X1)   integer part of X1.

The last function listed produces the integer 5 if, for example, 5.75 is in X1. It produces 5 if 5.0 is in X1, and it produces -5 if -5.75 is in X1.

For the functions, the argument is placed in parentheses following the function name. The argument may be any valid PASCAL expression which can be evaluated as a number. Note that some of the examples in Fig. E-1 use a function as part of the argument of another function.

PASCAL has no exponentiation operation. Instead, one asks the machine to multiply the natural logarithm of the number by the power, then use the result as the exponent of e. The original number cannot be negative. The one exception is that SQR(X1) can be used to square X1 or another number, used as its argument.

A number of programs in this book use a function to be supplied by the user at the time the machine is programmed. A typical flow chart statement is f(X1,X2)→X7. When this occurs, the novice programmer should simply substitute the correct PASCAL expression so that the machine evaluates the function and places its value in the proper memory location. More advanced programmers should learn to program PASCAL "procedures" and "functions". See the instruction manual for details.

Output can be obtained by means of a statement of the form

$$WRITELN(X1,X2,X3)$$

which tells the machine to write a line of output, containing the values of the variables enclosed by the parentheses. Some standard format, peculiar to the installation, is used.

Unconditional transfer is accomplished by labelling a statement with an integer number, then using a GO TO instruction to tell the machine to go to the labelled statement. Label numbers must have 4 or less digits, occur before the statement they label, and be followed by a colon. Statement 50, for example, might be

$$50:X1:=X2+X3$$

Somewhere else in the program the statement

$$GO\ TO\ 50$$

might appear. When the machine encounters this, it immediately executes statement 50 and succeeding statements, in order.

If any statements in the program are labelled, the labels must be delcared in a statement of the form

$$LABEL\ 50,720,818$$

where the numbers are statement labels used in the program, separated by commas. This declaration is placed after the program name declaration and before the variable declaration statement.

A possible conditional transfer statement has the form

```
IF X1<X2
    THEN GO TO 50;
X1:=X2+X3;
```

If the condition is true (if X1 has a value less than X2, in this case), then the machine goes to statement 50 and begins executing instructions there. If, on the other hand, the condition is false (if X1 is greater than or equal to X2), then the machine next executes the instruction following the THEN GO TO 50 line. In this case, it adds X2 and X3 and places the result in X1. There is no semicolon after IF X1<X2.

Other comparison statements which can be used are

	X1=X2	true if X1 equals X2,
	X1<>X2	true if X1 does not equal X2,
	X1>X2	true if X1 is greater than X2,
	X1>=X2	true if X1 is greater than or equal to X2,
and	X1<=X2	true if X1 is less than or equal to X2.

Any PASCAL arithmetic expressions which can be evaluated as numbers can be compared in the comparison part of the statement. For example,

```
IF X1*X2>X1+X2
    THEN ...
```

is valid provided values have previously been stored in X1 and X2.

The statement following THEN may be any valid PASCAL statement. Examples are

```
        THEN WRITELN(X1);
```

and   `THEN X1:=X2+X3;`

If the comparison condition is true, the statement is executed and then the machine proceeds to the following instruction. If the comparison condition is false, the machine ignores the line beginning with THEN and proceeds immediately to the following line.

PASCAL language contains statements which allow the construction of loops in more efficient ways than with the use of IF ... THEN instructions. We give one method. The appropriate sequence has the form

```
FOR I:=15 TO 32 DO
    BEGIN
        X1:=(I-1)*X2;
        X3:=SIN(X1);
        WRITELN(X3)
    END;
```

Statements in the loop are programmed between the two words BEGIN and END. If there is only one statement in the loop, the words BEGIN and END may be omitted. When there is more than one statement in the loop, the statements are separated by semicolons, with no semicolon after BEGIN and no semicolon before END.

For the example above, statements in the loop are first executed with the number 15 stored in I and, on successive executions of the loop, the value of I is incremented by unity. The last time through the loop, I has the value 32. I need not appear in any of the statements of the loop. If it does not, it is used simply to count the number of times the loop statements are executed.

I must be declared an integer variable in the variable declaration statement. The initial and final values (15 and 32, respectively, in the example) may be any integers with the final value greater than the initial value. They may also be PASCAL expressions containing integers and declared integer variables.

There are other PASCAL statements which can be used to construct loops. The reader is encouraged to learn about them from the instruction manual.

Fig. E-2 shows a PASCAL program for the calculation described in the introduction to the appendices. The first line gives the program name (SUM) and indicates that the

Figure E-2.   Sample program.

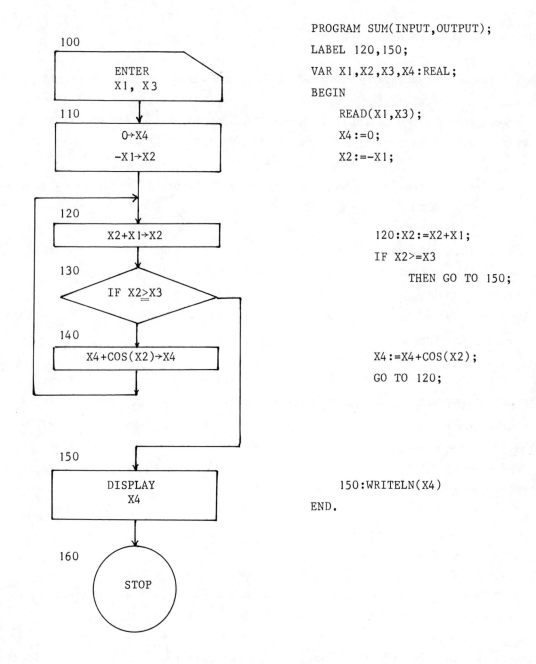

```
PROGRAM SUM(INPUT,OUTPUT);
LABEL 120,150;
VAR X1,X2,X3,X4:REAL;
BEGIN
    READ(X1,X3);
    X4:=0;
    X2:=-X1;


    120:X2:=X2+X1;
    IF X2>=X3
        THEN GO TO 150;


    X4:=X4+COS(X2);
    GO TO 120;



    150:WRITELN(X4)
END.
```

program will ask the machine to accept input data and to write lines of output. Statement numbers (120 and 150) are declared next. In the next line, variables are declared. The program uses four variables, named X1, X2, X3, and X4. All are real (not integer). The order of the declaration statements is important. It is also important that all of these statements are followed by semicolons.

The main body of the program is placed between the words BEGIN and END. END is followed by a period, indicating the end of the program. Each statement is followed by a semicolon, except there is no semicolon after BEGIN or before END. IF X2>=X3 THEN GO TO 150 is a single, simple statement and no semicolon breaks it.

All rules discussed previously have been followed. Carefully compare the program with the flow chart statements.

For some programs in this book, the memory locations for some of the variables must be computed by the machine during the running of the program. In the flow chart, we might write, for example, X(3*X2+1), X(3*X2+2), and X(3*X2+3) for three of the variables. These are called subscripted variables and the number between the parentheses, which gives the memory location number, is called the subscript.

PASCAL allows for subscripted variables. They are written X[I] and are used, just as any other variable is used, in arithmetic statements or as arguments of function. Note the square brackets. The subscript I must have a numerical value when the machine encounters the variable and the subscript must be of type integer. Storage of an input number in memory 3*X2+1, where X2 is a real variable, is accomplished by means of the following steps:

```
        I:=TRUNC(3*X2+1);
        READ(X[I])
```

TRUNC(X1) is a function operation which converts a real variable, in the argument, to an integer variable, truncating the value of the variable, if necessary.

Subscripted variables must be declared in the variable declaration statement after the program name and label declaration statements. The form is

         VAR X:ARRAY [1..50] OF REAL;
             X1:REAL;
             I:INTEGER;

First comes the name of the variable, X, then a colon, followed by the word ARRAY to tell the machine the variable is subscripted. The numbers in the square brackets give the minimum and maximum value of the subscript, separated by two periods. We arbitrarily picked I to run from 1 to 50, inclusive. The words OF REAL indicate that the variables X[I] are of type real. This is replaced by OF INTEGER if the X's are of type integer. The two succeeding lines, indicating that X1 is a real variable and I is an integer variable, are written to remind you that all three types are part of a single variable declaration statement. The word VAR appears only once.

/530H188FSUPP.>C1/